EAST MIDLANDS GEOTECHNICAL GROUP
THE INSTITUTION OF CIVIL ENGINEERS

Problematic Soils

Proceedings of the symposium
held at The Nottingham Trent University
School of Property and Construction
on 8 November 2001

Editors: I. Jefferson, E.J. Murray, E. Faragher and
P.R. Fleming

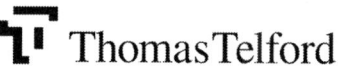 ThomasTelford

Organizers: The East Midlands Geotechnical Group of the Institution of Civil Engineers

Organizing committee: Dr I. Jefferson, School of Property and Construction, The Nottingham Trent University; Dr E.J. Murray, Murray Rix Geotechnical; E. Faragher, Advantica Technologies Ltd., Dr P.R. Fleming, Department of Civil and Building Engineering, Loughborough University.

Published for the organizers by Thomas Telford Publishing, Thomas Telford Ltd, 1 Heron Quay, London E14 4JD. URL: http://www.thomastelford.com

Distributors for Thomas Telford books are
USA: ASCE Press, 1801 Alexander Bell Drive, Reston, VA 20191-4400, USA
Japan: Maruzen Co. Ltd, Book Department, 3–10 Nihonbashi 2-chome, Chuo-ku, Tokyo 103
Australia: DA Books and Journals, 648 Whitehorse Road, Mitcham 3132, Victoria

First published 2001

A catalogue record for this book is available from the British Library

ISBN: 0 7277 3043 6

Printed and bound in Great Britain by MPG Books, Bodmin, Cornwall illustrates their behaviour with specific reference to the case of a very soft oragnic silt in Limerick, Ireland. Wilson provides a similar illustration of the behaviour of peat in his discussion of the peats from the Somerset Levels. The final paper that deals with naturally occurring problematic deposit is that of Cripps and Czerewko, who present work on the assessment of problem mudrocks, which are frequently encountered in the UK.

The next topic covered is that of problems associated with man-made or man-altered soils. Card and Richards examine issues related to building on landfill, with specific reference to policy and sources of information available as well as research currently being conducted. Bunce and Braithwaite provide details of a recent case history (Pride Park, Derby) where contaminated land has been successfully reclaimed. Skinner examines the behaviour and engineering characteristics of construction fill, an aspect of growing importance with the increase in development of brownfield sites which typically consist of substantial depths of fill. Finally, Coombs et al., consider whether it is the soil that is the problem or whether it is actually that specification and assessment criteria that have not been developed sufficiently to deal with those soils traditionally considered as problematic. Specific emphasis is given for the use of recycled materials, and the problems are expected to become increasingly important as the economic and political pressure to use these materials mounts.

The final area covered by the symposium deals with the treatment of problematic soils. This includes dealing with quicksand and other dewatering problems (Preene), deep compaction techniques (Slocombe) and the use of cement to stabilise contaminated clay (Maries et al.).

The final paper in this symposium is the keynote paper presented by the 1999 Rankine lecturer, Professor Serge Leroueil. His paper "No problematic soils, only engineering solutions," reconsiders the original definition of what a problematic soils is and suggests that by using a logical framework of observation, understanding and technology (OUT) soils can be engineered safely and economically. He postulates there are no problematic soils, just soils that require careful consideration to successfully engineer. Such an approach has, perhaps, the potential to save much of the billions of pounds spent globally each year, in the remediation of these soils

Although there have been a number of conferences, meetings and books published throughout the world on problematic soils, it was felt by the organizing committee that many of these were aimed at specific groups of practitioners and researchers. Moreover, it was felt that there has been insufficient focus on recent UK problems, development and practices. It was therefore considered timely to hold a symposium that covered a range of issues relevant to engineers with the aim of highlighting new areas of research and practice, and to provide a forum for discussion of these. The East Midlands Geotechnical Group (EMGG) therefore decided to organize a symposium on this subject in 2001, following their successful seminars on Groundwater Pollution in 1994, Lime Stabilization in 1996 and Geotechnical Engineering of Landfill 1998. This symposium was also supported by the British Geological

Preface

The symposium on Problematic Soils and this associated publication aim to provide an opportunity for dissemination of current practice and discussion on recent developments in the geotechnical engineering of problematic soils. Specific objectives are:

♦ to highlight the importance of understanding the geology and geomorphology;
♦ to identify and address the difficult ground conditions encountered;
♦ to present methods (through case studies) of *in situ* treatment of problematic soils, and
♦ to develop a sounder engineering approach for problematic soils.

In organizing this symposium, the editors considered that problematic soils are those which are difficult (or problematic) for the engineer to deal with. These soils have been, and continue to be a major geohazard to the built environment, with billions of pounds spent the world over preventing and remediating damage suffered as a result of their behaviour. This symposium aims to bring together in one volume a collection of papers that highlights the behaviour and characteristics of problematic soils (particularly those found in the UK), and offers guidance on possible treatment techniques that could be applied for their successful engineering. However, ground related problems are numerous with many different type of geological deposit presenting interesting and sometimes difficult challenges. It is for these reasons that this symposium focuses on soils, particularly those that are fine-grained. The proceedings of this symposium comprises three sections. The first section deals with naturally occurring problematic soils. The second section deals with man-made/altered problematic soils, an area that is becoming increasingly important. The final section deals with the engineering treatment of problematic soils. The symposium is finished off with the keynote paper "No problematic soils, only engineering solutions," by Professor Leroueil of the Université Laval, Canada.

The section on naturally occurring problematic soils starts with an excellent review of a range of problem soils found in the UK by Bell and Culshaw. They review the geology, behaviour and treatment of swelling/shrinking clays, collapsible soils, quicksands, frozen soils, peat and soils altered by weathering. These are then dealt with in more detail within the five subsequent papers. Jefferson *et al.,* discuss collapsible loess soils (more commonly known as brickearth) in the UK, highlighting their occurrence and engineering behaviour and illustrated with a case history. Further details of the problems associated with the swelling/shrinkage behaviour of clays are presented by Driscoll and Chown, who provide possible options to deal with them, including alternative foundation and remedial measures. Following on this is a discussion of the behaviour of highly compressible clays and silts is given by Farrell, who

Contents

Problem soils: a review from a British Perspective

F.G. Bell and M.G. Culshaw
British Geological Survey, Nottingham, NG12 5GG

Introduction

Unfortunately, many soils can prove problematic in geotechnical engineering, because they expand, collapse, disperse, undergo excessive settlement, have a distinct lack of strength or are corrosive. Such characteristics may be attributable to their composition, the nature of their pore fluids, their mineralogy or their fabric. Soil, with the exception of peat, is formed by the breakdown of rock masses either by weathering or erosion. It may accumulate in place or undergo a certain amount of transport, either of which influences its character and behaviour. However, the type of breakdown suffered by rock masses, and therefore the resulting soils, can be influenced profoundly by the climatic regime in which they are developed. The stage of maturity that a soil has reached also influences its behaviour.

There are many types of problem soils, some of the most noteworthy being swelling/shrinking clays, collapsible soils, quick sands, frozen soils and peat. The consequences that may be attributable to the behaviour of such problem soils can result in considerable financial loss. For instance, in the last decade, or so, swelling and shrinkage in clay soils have caused losses of up to £3 billion in Britain.

In Britain one has to bear in mind that until the recent geological past, much of the country was covered by ice, which either removed soil or left behind its own characteristic deposits. South of the ice masses ground was frozen, which also has left its effects. Nonetheless, geology does play its part, in that many formations, notably from the Jurassic onwards, contain deposits that have presented problems to engineers working with them. In this context, yet another factor must be taken into account and that is weathering. For example, not only does weathering reduce the strength of clay soils but it also facilitates the development of fissures. The latter play an extremely important role in the failure mechanism of fissured clays.

Problematic Soils. Thomas Telford, London, 2001

Expansive clays

Some clay soils undergo slow volume changes that occur independently of loading and are attributable to swelling or shrinkage. These volume changes can give rise to ground movements which can cause damage to buildings. Low-rise buildings are particularly vulnerable to such ground movements since they generally do not have sufficient weight or strength to resist. In addition, shrinkage settlement of embankments can lead to cracking and break up of the roads they support.

The principal cause of expansive clays is the presence of swelling clay minerals such as montmorillonite. Differences in the period and amount of precipitation and evapotranspiration are the principal factors influencing the swell-shrink response of a clay soil beneath a building. Poor surface drainage or leakage from underground pipes also can produce concentrations of moisture in clay. Trees with high water demand and uninsulated hot process foundations may dry out clay causing shrinkage.

The depth of the active zone in expansive clays (i.e. the zone in which swelling and shrinkage occurs in wet and dry seasons respectively) varies. Many soils in temperate regions such as Britain, especially in the south, south east and south Midlands of England, possess the potential for significant volume change due to changes in moisture content. However, owing to the damp climate in most years volume changes are restricted to the upper 1.0 to 1.5 m in clay soils. The susceptible soils include most of the clay formations of Mesozoic and Tertiary age such as the Edwalton Formation of the Mercia Mudstone, Lias Clay, Fullers' Earth, Blisworth Clay, Kellaways Clay, Oxford Clay, Ampthill Clay, Kimmeridge Clay, Wadhurst Clay, Weald Clay, Atherfield Clay, Gault Clay, clays of the Lambeth Group, London Clay and Claygate Beds and Barton Clay. Weathered Namurian and Coal Measures shales and some glacial tills derived from any of the clay and shale formations also may be susceptible. Generally speaking, the older formations are less susceptible than the younger ones.

The potential for volume change in clay soil is governed by its initial moisture content, initial density or void ratio, its microstructure and the vertical stress, as well as the type and amount of clay minerals present. These clay minerals are responsible primarily for the intrinsic expansiveness whilst the change in moisture content or suction (where the pore water pressure in the soil is negative, that is, there is a water deficit in the soil) controls the actual amount of volume change, which a soil undergoes at a given applied pressure. The rate of heave depends upon the rate of accumulation of moisture in the soil.

Grim (1962) distinguished two modes of swelling in clay soils, namely, intercrystalline and intracrystalline swelling. Interparticle swelling takes place in any type of clay deposit irrespective of its mineralogical composition, and the process is reversible. In relatively dry clays the particles are held together by relict water under tension from capillary forces. On wetting, the capillary force is relaxed and the clay expands. In other words, intercrystalline swelling takes place when the uptake of moisture is restricted to the external crystal surfaces and the void spaces between the crystals. Intracrystalline swelling, on the other hand, is characteristic of the smectite family of clay minerals, and montmorillonite in particular. The individual molecular layers, which make up a crystal of montmorillonite are weakly bonded so that on wetting water enters not only between the crystals but also between the unit layers which comprise the crystals. Generally kaolinite has the smallest swelling capacity of the clay minerals and nearly all of its swelling is of the interparticle type. Illite may swell by up to 15% but intermixed illite and montmorillonite may swell some 60 to 100%. Swelling in calcium montmorillonite is very much less than in the sodium variety, it ranging from about 50 to 100%. Swelling in sodium montmorillonite can amount to 2 000% of the original volume, the clay then having formed a gel.

Cemented and undisturbed expansive clay soils often have a high resistance to deformation and may be able to absorb significant amounts of swelling pressure. Therefore, remoulded expansive clays tend to swell more than their undisturbed counterparts. In less dense soils expansion initially takes place into zones of looser soil before volume increase occurs. However, in densely packed soil with low void space, the soil mass has to swell more or less immediately to accommodate the volume change. Therefore, clay soils with a flocculated fabric swell more than those that possess a preferred orientation. In the latter, the maximum swelling occurs normal to the direction of clay particle orientation. Because expansive clays normally possess extremely low permeabilities, moisture movement is slow and a significant period of time may be involved in the swelling-shrinking process. Accordingly, moderately expansive clays with a smaller potential to swell but with higher permeabilities than clays having a greater swell potential may swell more during a single wet season than more expansive clays.

The swell-shrink behaviour of a clay soil under a given state of applied stress in the ground is controlled by changes in soil suction. The relationship between soil suction and water content depends on the proportion and type of clay minerals present, their microstructural arrangement and the chemistry of the pore water. Changes in soil suction are brought about by moisture movement through the soil due to evaporation from its surface in dry weather, by transpiration from plants, or alternatively by recharge consequent upon precipitation. The climate governs the amount of moisture available to counteract that, which is removed by evapotranspiration (i.e. the soil moisture deficit). The volume changes that occur due to evapotranspiration from clay soils can be conservatively predicted by assuming the lower limit of the soil moisture content to be the shrinkage limit. Desiccation beyond this value cannot bring about further volume change.

Transpiration from vegetative cover can represent an important means of water loss from soils in dry summers. Indeed, the distribution of soil suction in soil may be controlled by transpiration from vegetation and represents one of the significant changes made in loading (i.e. to the state of stress in a soil). The maximum soil suction that can be developed is governed by the ability of vegetation to extract moisture from the soil. The level at which moisture is no longer available to plants is termed the permanent wilting point and this corresponds to a pF value of about 4.2. The pF index value is a soil suction index quantity related to negative capillary suction. The moisture characteristic (moisture content *v.* soil suction) of a soil provides valuable data concerning the moisture contents corresponding to the field capacity (defined in terms of soil suction this is a pF value of about 2.0) and the permanent wilting point, as well as the rate at which changes in soil suction take place with variations in moisture content. This enables an assessment to be made of the range of soil suction and moisture content which is likely to occur in the zone affected by seasonal changes in climate. The suction pressure associated with the onset of cracking is approximately pF 4.6.

The extent to which vegetation is able to increase the suction to the level associated with the shrinkage limit obviously is important. In fact, the moisture content at the permanent wilting point exceeds that of the shrinkage limit in soils with high contents of expansive clay minerals and is less in those possessing low contents. This explains why settlement resulting from the desiccating effects of trees is more notable in low to moderately expansive soils than in expansive ones.

Many clay soils in Britain possess the potential for volume change but the climate means that any significant deficits in soil moisture developed during the summer are confined to the upper 1.0 to 1.5 m of the soil and the field capacity is re-established during the winter. Nonetheless, deeper permanent deficits can be brought about by large trees. With this in mind, Driscoll (1983) suggested that desiccation could be regarded as commencing when the rate of change in moisture content (and therefore volume) with increasing soil suction increases significantly. He proposed that this point approximates to a suction of pF about 2 (10 kPa). Similarly, notable suction could be assumed to have taken place if, on its disappearance, a low-rise building was uplifted due to the soil swelling. Driscoll maintained that this suction would have a pF value of 3 (100 kPa).

As far as *in situ* testing is concerned, initial effective stresses can be estimated with the aid of a psychrometer used to measure soil suction. Gourley *et al.,* (1994) referred to the use of a suction probe for measuring soil suction in the field. In addition, settlement points can be installed at different depths in the ground using sleeved rods to measure the seasonal movements of the soil in conjunction with moisture content. These measurements provide direct evidence of potential shrinkage and swelling movements. However, such measurements are time consuming. In addition, Williams and Pidgeon (1983), admittedly in the context of South Africa, pointed out that the measurement of seasonal ground movements under natural conditions might give appreciable underestimates of potential total or differential movements under buildings, particularly when desiccation extends to some depth.

Methods of predicting volume changes in soils can be grouped into empirical methods, soil suction methods and oedometer methods (Bell and Maud, 1995). Empirical methods make use of the swelling potential as determined from void ratio, natural moisture content, liquid and plastic limits, and activity. For example, Driscoll (1983) proposed that the moisture content, m, at the onset of desiccation (pF = 2) and when it becomes significant (pF = 3) could be approximately related to the liquid limit, LL, in the first instance m = 0.5 LL and in the second m = 0.4 LL. The Building Research Establishment (BRE) (Anon, 1980) suggested that the plasticity index provided an indication of volume change potential as shown in Table 1. A degree of overlap was allowed. The activity chart, proposed by Van der Merwe (1964), frequently has been used to assess the expansiveness of clay soils (Figure 1).

Plasticity Index (%)	Potential for volume change
Over 35	Very high
22-48	High
12-32	Medium
Less than 18	Low

Table 1 Soil activity related to swelling.

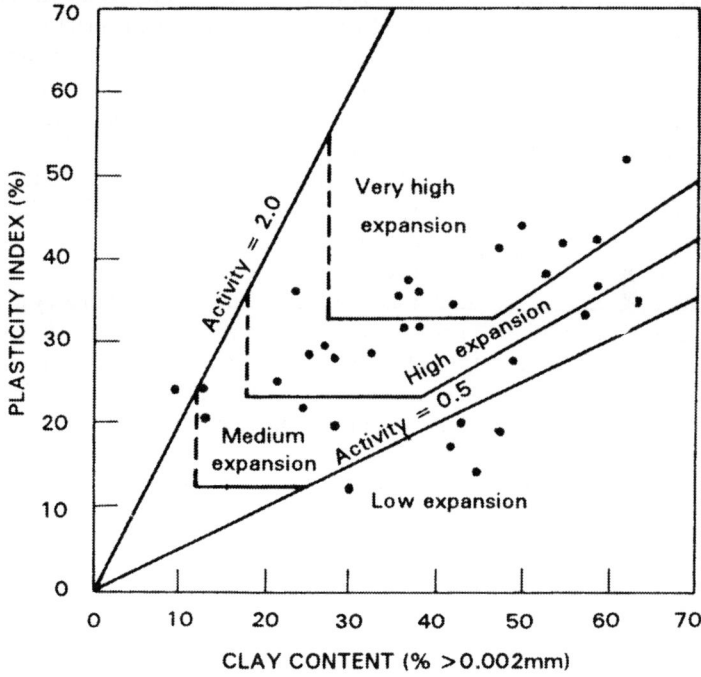

Figure 1 Activity chart of Van der Merwe (1964) for estimation of the degree of expansiveness of clay soil. Some expansive clays from Natal, South Africa, are shown (after Bell and Maud, 1995).

However, because the determination of plasticity is carried out on remoulded soil, it does not consider the influence of soil texture, moisture content, soil suction or pore water chemistry, which are important factors in relation to volume change potential. Therefore, over-reliance on the results of such tests must be avoided. Consequently, empirical methods should be regarded as simple swelling indicator methods and nothing more. As such, it is wise to carry out another type of test and to compare the results before drawing any conclusions.

Soil suction methods use the change in suction from initial to final conditions to obtain the degree of volume change. Soil suction is the stress which, when removed allows the soil to swell, so that the value of soil suction in a saturated fully-swollen soil is zero. O'Neill and Poormoayed (1980) quoted the United States Army Engineers Waterways Experimental Station (USAEWES) classification of potential swell (Table 2) which is based on the liquid limit, plasticity index and initial (*in situ*) suction. The latter is measured in the field by a psychrometer. Soil suction is not easy to measure accurately. Filter paper has been used for this purpose (McQueen and Miller, 1968). According to Chandler *et al.,* (1992), measurements of soil suction obtained by the filter paper method compare favourably with measurements obtained using psychrometers or pressure plates.

Liquid limit (%)	Plastic limit (%)	Initial (*in situ*) suction (kPa)	Potential swell (%)	Classification
Less than 50	Less than 25	Less than 145	Less than 0.5	Low
50-60	25-35	145-385	0.5-1.5	Marginal
Over 60	Over 35	Over 385	Over 1.5	High

Table 2 USAEWES classification of swell potential (from O'Neill and Poorymoayed, 1980).

The oedometer methods of determining the potential expansiveness of clay soils represent more direct methods (Jennings and Knight, 1957). In the oedometer methods, undisturbed samples are placed in the oedometer and a wide range of testing procedures are used to estimate the likely vertical strain due to wetting under vertical applied pressures. The latter may be equated to overburden pressure plus that of the structure which is to be erected. In reality most expansive clays are fissured, which means that lateral and vertical strains develop locally within the ground. Even when the soil is intact, swelling or shrinkage is not truly one-dimensional. The effect of imposing zero lateral strain in the oedometer is likely to give rise to overpredictions of heave and the greater the degree of fissuring, the greater the overprediction. The values of heave predicted using oedometer methods correspond to specific values of natural moisture content and void ratio of the sample. Therefore, any change in these affects the amount of heave predicted. Gourley *et al.*, (1994) mentioned the use of a stress path oedometer to determine volume change characteristics of expansive soils. Such a method provides data on vertical and radial total stresses, suction and void ratio.

Effective and economic foundations for low-rise buildings on swelling and shrinking soils have proved difficult to achieve. This is partly because the cost margins on individual buildings are low. Obviously, detailed site investigation and soil testing are out of the question for individual dwellings. Similarly, many foundation solutions that are appropriate for major structures are too costly for small buildings. Nonetheless, the choice of foundation is influenced by the subsoil and site conditions, estimates of the amount of ground movement and the cost of alternative designs. In addition, different building materials have different tolerances to deflections (Burland and Wroth, 1975). Hence, materials that are more flexible can be used to reduce potential damage due to differential movement of the structure.

Three methods can be adopted when choosing a design solution for building on expansive soils, namely, provide a foundation and structure which can tolerate movements without unacceptable damage; isolate the foundation and structure from the effects of the soil; or alter or control the ground conditions. In addition to these construction details, moisture control measures should be adopted as far as possible.

The isolation of foundation and structure has been widely adopted for 'severe' and 'very severe' ground conditions. Straight-shafted bored piles can be used in conjunction with suspended floors for severe conditions. The piles are sleeved over the upper part and provided with reinforcement. For severe conditions it may be necessary to place piles at appreciable depth (i.e. below the level of fluctuation of natural moisture content) and/or use under-reams to resist the pull-out forces.

The use of stiffened rafts is fairly commonplace (Bell and Maud, 1995). The design of the slab is dependent on assumed allowable relative deflections and it is usually necessary to incorporate anti-cracking features in the superstructure such as flexible joints.

Moisture control is perhaps the most important single factor in the success of foundations on shrinking and swelling clays. The aim is to maintain stable moisture conditions with minimum moisture content or suction gradients. The loss of moisture around the edges of a building, which leads to the moisture content of the soil under the centre of the building being higher, gives rise to differential heave. In order to control this an attempt should be made to maintain the same moisture content beneath a building. This can be achieved by the use of horizontal and vertical moisture barriers around the perimeter of the building, drainage systems and control of vegetation coverage.

A simple method of reducing or eliminating ground movements due to expansive soil is to replace or partially replace them with non-expansive soils.

There is no requirement for the thickness of the replacement material but a minimum of 1 m has been suggested by Chen (1988). The material should be granular but it should not allow surface water to travel freely through the soil so that it wets any swelling soils in lower horizons. Therefore, the presence of a fine fraction is required to reduce permeability or, better still, a geomembrane can surround the granular material.

If expansive soil is allowed to swell by wetting prior to construction and if the soil moisture content then is maintained, the soil volume should remain relatively constant and no heave should take place. Ponding is the most common method of wetting. This may take several months to increase the water content to the required depth, notably in areas with deep groundwater surfaces. Vertical wells can be installed to facilitate flooding and thus decrease the time necessary to adjust the moisture content of the soil.

The amount of heave of expansive soils is reduced significantly when compacted to low densities at high moisture contents. Expansive soils compacted above optimum moisture content undergo negligible swell for any degree of compaction. On the other hand, compaction below optimum results in excessive swell.

Many attempts have been made to reduce the expansiveness of clay soil by chemical stabilization. For example, lime stabilization of expansive soils, prior to construction, can minimize the amount of shrinkage and swelling they undergo. In the case of light structures, lime stabilization may be applied immediately below strip footings. However, significant sulphate (SO_4) content (i.e. in excess of 5 000 mg kg^{-1}) in clay soils can mean that they react with calcium oxide (CaO) to form ettringite with resultant expansion. The treatment is better applied as a layer beneath a raft so as to overcome differential movement. The lime stabilized layer is formed by mixing 4 to 6% lime with the soil. A compacted layer, 150 mm in thickness, usually gives satisfactory performance. Furthermore, the lime stabilized layer redistributes unequal moisture stresses in the subsoil so minimizing the risk of cracking in the structure above, as well as reducing water penetration beneath the raft. Alternatively, lime treatment can be used to form a vertical cut-off wall at, or near, the footings in order to minimize movement of moisture.

Cement stabilization has much the same effect on expansive soils as lime treatment, although the dosage of cement needs to be greater for heavy expansive clays. Alternatively, they can be pretreated with lime, thereby reducing the amount of cement that needs to be used.

Collapsible soils

Soils such as loess, brickearth and certain wind blown silts may possess the potential to collapse. These soils generally consist of 50 to 90% silt particles and sandy, silty and clayey types have been recognized by Clevenger (1958), with most falling into the silty category. Collapsible soils possess porous textures with high void ratios and relatively low densities. They often have sufficient void space in their natural state to hold their liquid limit moisture at saturation. At their natural low moisture content these soils possess high apparent strength but they are susceptible to large reductions in void ratio upon wetting. In other words, the metastable texture collapses as the bonds between the grains break down when the soil is wetted. Hence, the collapse process represents a rearrangement of soil particles into a denser state of packing. Collapse on saturation normally only takes a short period of time, although the more clay such a soil contains, the longer the period tends to be.

The fabric of collapsible soils generally takes the form of a loose skeleton of grains (generally quartz) and microaggregates (assemblages of clay or clay and silty clay particles). These tend to be separate from each other, being connected by bonds and bridges, with uniformly distributed pores. The bridges are formed of clay-sized minerals, consisting of clay minerals, fine quartz, feldspar or calcite. Surface coatings of clay minerals may be present on coarser grains. Silica and iron oxide may be concentrated as cement at grain contacts and amorphous overgrowths of silica occur on grains. As grains are not in contact, mechanical behaviour is governed by the structure and quality of bonds and bridges. The structural stability of collapsible soils is not only related to the origin of the material, to its mode of transport and depositional environment but also to the amount of weathering undergone.

Popescu (1986) maintained that there is a limiting value of pressure, defined as the collapse pressure, beyond which deformation of soil increases appreciably. The collapse pressure varies with the degree of saturation. He defined truly collapsible soils as those in which the collapse pressure is less than the overburden pressure. In other words, such soils collapse when saturated since the soil fabric cannot support the weight of the overburden. When the saturation collapse pressure exceeds the overburden pressure soils are capable of supporting a certain level of stress on saturation and Popescu defined these soils as conditionally collapsible soils. Conditionally collapsible soils therefore can be described as psuedo-stable. The maximum load that such soils can support is the difference between the saturation collapse and overburden pressures. Phien-wej *et al.,* (1992) concluded that the critical pressure at which collapse of the soil fabric begins was greater in soils with smaller moisture content. Under the lowest natural moisture content the soils investigated posed a severe problem on wetting (the collapse-potential was as high as 12.5% at 5% natural moisture content). During the wet season when the natural moisture content could rise to 12%, there was a reduction in the collapse potential to around 4%.

In Britain brickearth, which occurs mainly in south east England, notably in Essex, Kent, Sussex, Hampshire and Dorset, is similar to loess. Small patches are found also in Devon (where it is described as 'Loam'). Similar material was also found in a sinkhole on the route of the M25 motorway north west of London (Gibbard, P. L., *pers. comm.*). On geological maps brickearth is named and described in a number of ways. These are shown in Table 3.

Name	Description
Brickearth	Varies from silt to clay, usually yellow-brown and massive
Head Brickearth	Varies from silt to clay, usually yellow-brown and massive. Poorly sorted and poorly stratified, formed mostly by solifluction and/or hillwash and soil creep
River Brickearth	Varies from silt to clay, usually yellow-brown and massive; of fluvial origin
Head Brickearth, Older	Varies from silt to clay, usually yellow-brown and massive. Poorly sorted and poorly stratified, formed mostly by solifluction and/or hillwash and soil creep; older than 'Head Brickearth' in the same map area
Head Brickearth, Younger	Varies from silt to clay, usually yellow-brown and massive. Poorly sorted and poorly stratified, formed mostly by solifluction and/or hillwash and soil creep. Younger than 'Head Brickearth' in the same map area

Table 3 Types of brickearth shown on British Geological Survey maps and listed in the BGS Lexicon of Rock Names.

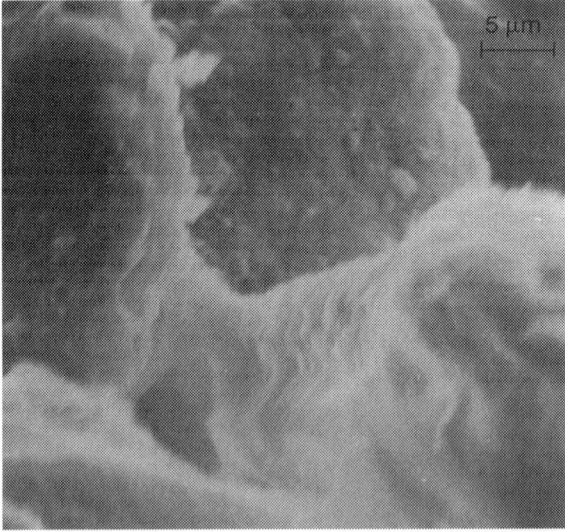

Figure 2 Scanning electron photomicrograph showing three silt particles, two bridged by clay particles, from brickearth of south Essex (after Northmore *et al.,* 1996).

Brickearth is composed largely of silt-sized particles. Generally, brickearth comprises a blanket deposit of yellowish brown, friable, slightly plastic, poorly bedded clayey and sandy silt with well developed vertical jointing. It has a very open, low density structure. Figure 2 shows three silt-size particles with two of them bridged by platy clay particles. Brickearth generally is calcareous at depth, although the upper parts, and often the full thickness, may be leached of carbonate material. The lower parts of a deposit tend to be more rigid and better consolidated than the upper parts.

Property	Maximum value	Minimum value	Mean
Specific gravity	2.77	2.62	2.70
Bulk density (Mg/m³)	2.073	1.696	1.900
Dry density (Mg/m³)	1.743	1.487	1.617
Void ratio	0.819	0.565	0.678
Porosity	45	36	40
Natural moisture content (%)	22	13	18
Degree of saturation (%)	96	44	72
Liquid Limit (%)	54	30	35
Plastic Limit (%)	22	16	20
Plasticity Index (%)	32	9	15
Liquidity Index	0.17	-0.89	-0.18
Consistency Index	1.9	0.9	1.2
Calcium carbonate content (%)	16.5	0.16	10.5
Organic content (%)	1.0	0.1	0.42
Collapse index, i_c	0.90	0.02	0.41
Absolute index, i_{ac}	13.0	0.2	5.7
Coefficient of collapsibility, C_{col}	0.050	-0.004	0.011
Compression index	0.33	0.15	0.24
Coefficient of volume compressibility (m²/MN)	0.149	0.035	0.095
Coefficient of consolidation (m²/yr)	94.10	72.29	89.30
Coefficient of permeability (m/s) (10^{-8})*	34.5	4.2	8.8
Coefficient of permeability (m/s) (10^{-8})+	4.4	1.0	2.6

*Value of permeability calculated from the lowest loading in the oedometer test (that is, 0-107 kPa).+Value of permeability calculated from the highest loading in the oedometer test (that is, 429-858 kPa).
Note: The values of C_{col}, m_v (in (m² MN^{-1})) and c_v are those obtained from the highest loading applied in the oedometer test (that is, 429-858 kPa). Degree of compressibility, C_c, over 0.3, very high; 0.15-0.3, high; 0.075-0.15, medium; less than 0.075, low.

Table 4 Representative values of some basic soil properties of brickearths from south Essex. (after Northmore *et al,* 1996).

Quartz is the most abundant mineral in brickearth, ranging between 12 and 54% in composition, (with an average of 33%), followed by feldspar. Of the clay-type minerals, mica is generally more abundant than montmorillonite, which, in turn, is more abundant than illite and kaolinite. For example, Northmore *et al.* (1996) found that montmorillonite averaged some 15% in the brickearth of south Essex with a maximum of 39% and kaolinite constituted generally less than 5%. Calcium carbonate occurs as grains, thin tube infillings and as concretionary nodules. When present it tends to account for less than 10% of the soil.

Brickearth has a natural moisture content that usually varies from 12 to 25%. The soils are of variable plasticity, ranging from low to high plasticity but most are of low plasticity. The spread of liquid limits may range from 30 to 60% and while plastic limits range from about 20 to 25% (Tables 4 and 5). Most brickearth has negative liquidity indices, which indicates that it is in a fairly brittle condition.

The particle size distribution of brickearth from south Essex and Kent are shown in Figure 3, from which it can be seen that they are similar. Like loess, clayey, silty and sandy brickearth can be recognized. For instance, the clay content in the brickearth from south Essex ranges from 4 to 42%, with an average of 21%. This compares with an average silt content of 59%, silt being the most important size fraction in this brickearth. Brickearths vary from uniformly to well sorted soils, with most being well sorted. They tend to possess an excess of fines.

Most bulk densities fall within the range 1.7 and 2.1 Mg m^{-3} (Table 4), with most dry densities varying from 1.52 to 1.73 Mg m^{-3}. The relatively low densities of brickearths are reflected in high void ratios, indicative of a normally open microstructure. As such, they are potentially metastable in that their structure may collapse on wetting.

Location	G_s	ρ_d Mg/m^3	e	n %	LL %	PL %	PI %	CaCO$_3$ (%)
Pine Farm Quarry	2.70	1.482	0.82	45	32	22	10	14.0
Ford	2.70	1.490	0.81	45	34	19	15	12.7
Pegwell Bay	2.69	1.640	0.64	39	29	18	11	16.2
Pegwell Bay buried channel	2.69	1.736	0.55	36	33	20	13	
Reculver	2.68	1.623	0.65	39	33	20	13	6.0
Sturry	2.69	1.692	0.59	37	44	22	22	
Northfleet	2.70	1.605	0.68	41	32	19	13	9.4
Average	2.69	1.620	0.68	40	34	20	14	11.7

Average values for each location

G_s, specific gravity; ρ_d, dry density; e, void ratio; n, porosity; LL, liquid limit; PL, plastic limit; PI, plasticity index; CaCO$_3$, calcium carbonate.

Table 5 Representative values of some basic soil properties of brickearths from Kent (after Derbyshire and Mellors, 1988).

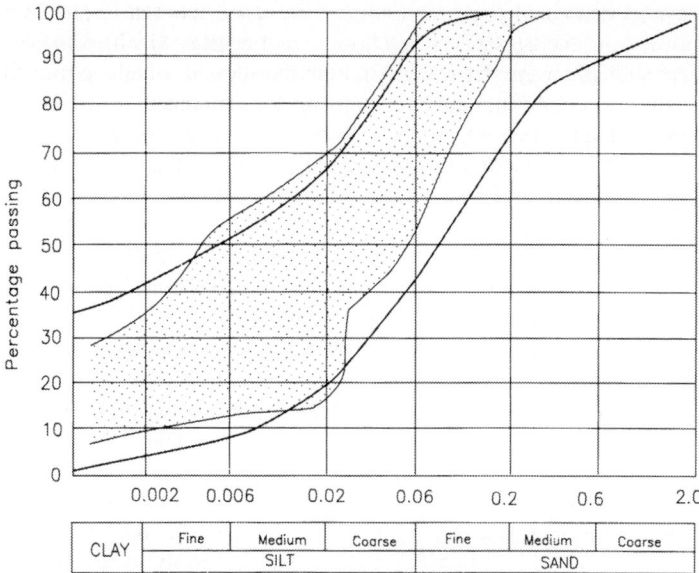

Figure 3 Particle size distribution of brickearth from south Essex compared with that from Kent (shown stippled), (after Northmore *et al.*, 1996).

The compression index indicates that the degree of compressibility of brickearth is high and generally the value of coefficient of volume compressibility decreases with increasing load. Tests on brickearth from south Essex carried out by Northmore *et al.*, (1996) exhibited rapid consolidation, which is reflected in the high coefficients of consolidation (Table 4). In many instances all the primary consolidation may take place within a half-minute of loading.

The results obtained from undrained triaxial tests on brickearth from south Essex by Northmore *et al.*, (1996) showed that the shear strength was between 10 and 220 kPa. Such a range indicates the variability in undrained shear strength. However, Northmore *et al.* noted a general tendency for shear strength to decrease with increasing depth. They suggested that this might be partly due to the variable composition of the deposit, which tends to vary from a stiff sandy silty clay (relatively dry) near the surface, to a clayey silty loam with increasing depth. The higher values of shear may reflect the 'crust-like' nature of the soil near the ground surface. Consolidated drained triaxial tests also indicate variable effective shear strengths. Peak values of internal angle of shearing resistance may be between 19° and 34°, and those of effective cohesion from between 10 to 70 kPa. Residual values drop to between 16° and 25°, and zero, respectively.

Northmore *et al.*, (1996) also investigated the effects of flooding on the undrained and drained shear strength parameters. They found that there was a dramatic decrease in undrained shear strength for the small increase in deviator stress. On removal from the triaxial cell, each specimen resembled a 'fluid paste'. The drained shear strengths were determined in both the shear box and the triaxial apparatus. A typical stress deformation relationship for specimens tested in the shear box is shown in Figure 4.

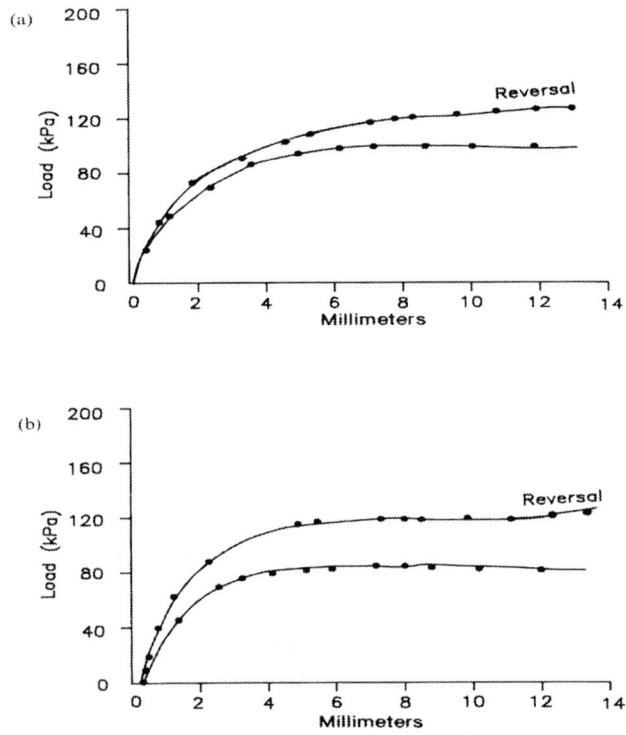

Figure 4 Typical drained shear box stress-strain curves for brickearth from south Essex (a) sheared at *in situ* moisture content (normal load = 100 kPa), (b) sheared in the saturated state (normal load = 100 kPa), (after Northmore *et al.,* 1996).

Figure 4 shows that the strength appears to increase dramatically (of the order of 20%) upon reversal of shear, suggesting that the soil matrix has been artificially disturbed during testing. Generally, stable and pseudo-stable samples did not exhibit significant changes in their shear strength characteristics due to flooding whereas, at least in the shear box apparatus, the metastable material exhibited a marked drop in effective shear strength following flooding. When tested under flooded conditions in the triaxial apparatus, the metastable soils showed a marked increase in peak angle of internal friction; the pseudo-stable samples demonstrated limited increase in peak angle of internal friction and the stable sample underwent no change in peak angle of internal friction. The effective cohesion of all the samples changed slightly upon flooding but not sufficiently to influence the shearing resistance of any of the soils at confining pressures in the engineering range.

Several collapse criteria have been proposed for predicting whether a soil is liable to collapse upon saturation. For instance, Clevenger (1958) suggested a criterion for collapsibility based on dry density, that is, if the dry density is less than 1.28 Mg m^{-3}, then the soil is liable to undergo significant settlement. On the other hand, if the dry density is greater than 1.44 Mg m^{-3}, then the amount of collapse should be small, while at intermediate densities the settlements are transitional. Gibbs and Bara (1962) suggested the use of dry unit weight and liquid limit as criteria to distinguish between collapsible and non-collapsible soil types. Their method is based on the premise that a soil, which has enough void space to hold its liquid limit moisture content at saturation, is susceptible to collapse on wetting. This criterion only applies if the soil is uncemented and the liquid limit is above 20%. When the liquidity index in such soils approaches or exceeds unity, then collapse may be imminent. As the clay content of a collapsible soil increases, the saturation moisture content becomes less than the liquid limit so that such deposits are relatively stable. However, Northmore *et al.,* (1996) concluded that this method did not provide a satisfactory means of identifying the potential metastability of brickearth. More simply, Handy (1973) suggested that collapsibility could be determined either by the percentage clay content or from the ratio of liquid limit to saturation moisture content. He maintained that soils with a:

- clay content of less than l6% had a high probability for collapse;
- clay content of between l6 and 24% were probably collapsible;
- clay content between 25 and 32% had a probability of collapse of less than 50%;
- clay content which exceeded 32% were non-collapsible.

Soils in which the ratio of liquid limit to saturation moisture content was less than unity were collapsible, while if it was greater than unity they were safe. After an investigation of loess in Poland, Grabowska-Olszewska (l988) suggested that loess with a natural moisture content less than 6% was potentially unstable, that in which the natural moisture content exceeds 19% could be regarded as stable, while that with values between these two figures exhibited intermediate behaviour.

Although, empirical indices of collapse may provide a relatively rapid and inexpensive means of determining collapsibility (Denisov, 1963; Feda, 1966; Fookes and Best, 1969; Derbyshire and Mellors, 1988; Northmore *et al.,* 1996), it must be borne in mind that some of the parameters, which are used for assessment are derived from remoulded soil samples. Such samples do not take account of the initial soil fabric. Consequently, the results should regarded as only general indicators of collapsibility and indeed some indices need to used with caution.

The oedometer test can be used to assess the degree of collapsibility. For example, Jennings and Knight (1975) developed the double oedometer test for assessing the response of a soil to wetting and loading at different stress levels (i.e. two oedometer tests are carried out on identical samples, one being tested at its natural moisture content, whilst the other is tested under saturated conditions, the same loading sequence being used in both cases). They subsequently modified the test so that it involved loading an undistributed specimen at natural moisture content in the oedometer up to a given load. At this point the specimen is flooded and the resulting collapse strain, if any, is recorded (Figure 5). Then the specimen is subjected to further loading.

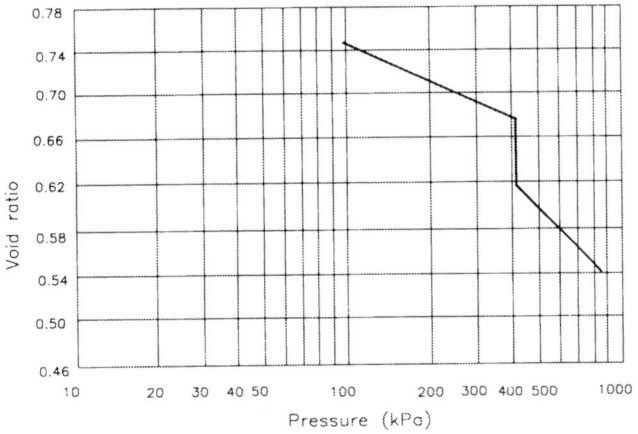

Figure 5 Void ratio-log pressure curve of metastable brickearth from south Essex tested in an oedometer (specimen flooded after 24 hours loading at 429 kPa), (after Northmore *et al.,* 1996).

Collapse (%)	Severity of problem
0-1	No problem
1-5	Moderate trouble
5-10	Trouble
10-20	Severe trouble
Above 20	Very severe trouble

Table 6 Collapse percentage (defined as $\Delta e/(1+e)$, were e is void ratio) as an indication of potential severity.(after Jennings and Knight, 1975).

Table 6 provides an indication of the potential severity of collapse. This Table indicates that those soils, which undergo more than 1% collapse can be regarded as metastable. However, in China a figure of 1.5% is taken (Lin and Wang, 1988) and in the United States values exceeding 2% are regarded as indicative of soils susceptible to collapse (Lutenegger and Hallberg, 1988).

In some parts of the world significant settlements beneath structures on collapsible soils have suffered foundation failure, especially after the soils have been wetted. Clemence and Finbarr (1981) recorded a number of techniques that could be used to stabilize collapsible soils. These are summarized in Table 7.

Various methods of compaction have been used to densify collapsible soils such as dynamic compaction and the use of compaction piles. However, if these soils contain a relatively high carbonate contents it may be difficult to achieve the desired result with dynamic compaction. Concrete compaction piles may be driven into the ground or compacted soil-cement piles have been employed. Gibbs and Bara (1967) referred to the use of clay grout, it being injected under pressure to compact loess. Compaction also has been brought about by vibration, either by deep vibroflotation or deep explosion. In both cases the soil has been wetted beforehand (Litvinov, 1973). Moreover, compaction has been achieved by inundation. Cement, lime and bitumen emulsion have been used to stabilize loess soils, especially in relation to road construction

Depth of treatment	Foundation treatment
0 - 1.4 m	A. Current and past methods Moistening and compaction (conventional extra heavy impact or vibratory rollers)
1.5 - 10 m	Over-excavation and recompaction (earth pads with or without stabilization by additives such as cement or lime). Vibro-flotation (free draining soils). Vibroreplacement (stone columns). Dynamic compaction. Compaction piles. Injection of lime. Lime piles and columns. Jet grouting. Ponding or flooding (if no impervious layer exists). Heat treatment to solidify the soils in place.
Over 10 m	Any of the aforementioned or combinations of the aforementioned, where applicable. Ponding and infiltration wells, or ponding and infiltration wells with the use of explosive. B. Possible future methods Ultrasonics to produce vibrations that will destroy the bonding mechanics of the soil. Electrochemical treatment. Grouting to fill pores.

Table 7 Methods of treating collapsible foundations (after Clemence and Finbarr, 1981).

Evstatiev (1988) referred to the use of soil or soil-cement cushions which replace loess, usually to a depth of 1.5 m and rarely reaching 3 m, and these are properly compacted in layers. Silicate grouts have been employed to treat loess soils in Russia and more recently jet grouting has proved an effective method of stabilization. Prevention of saturation around the foundations of small buildings, by using flexible drains and pipes, and by ensuring that the run-off is kept away by the use of concrete aprons, may be sufficient to avoid collapse taking place.

Peat soils

Peat represents an accumulation of partially decomposed and disintegrated plant remains that has been preserved under conditions of incomplete aeration and high water content. It accumulates wherever the conditions are suitable, that is, in areas where there is an excess of rainfall and the ground is poorly drained, irrespective of latitude or altitude. Nonetheless, peat deposits tend to be most common in those regions with a comparatively cold wet climate. Physicochemical and biochemical processes cause this organic material to remain in a state of preservation over a long period of time. In other words, waterlogged poorly drained conditions not only favour the growth of particular types of vegetation but they also help preserve the plant remains. The high water-holding capacity of peat maintains a surplus of water, which ensures continued plant growth and consequent peat accumulation. The rate of decomposition of plant detritus is relatively rapid at the surface where aerobic conditions exist but is slowed down several thousand-fold under anaerobic conditions. Drying out, groundwater fluctuations and snow loading bring about compression in the upper layers of a peat deposit. Indeed, these mechanisms are often more important, as far as near-surface compression is concerned, than effective overburden pressure. This is because the unit weight of peak may be similar to that of water. As the water table in peat normally is near the surface, the effective overburden pressure is negligible.

Macroscopically peaty material was divided by Radforth (1952) into three basic groups, namely, amorphous granular, coarse fibrous and fine fibrous peat. The amorphous granular peats have a high colloidal fraction, holding most of their water in an adsorbed rather than a free state, the adsorption occurring around the grain structure. Landva and Pheeney (1980), however, maintained that the term amorphous granular should be reserved for non-fibrous, truly amorphous peats only. They contended that most such material referred to in engineering literature is in fact moss peat, actual 'amorphous granular' peat being rare. They therefore suggested that the term should be used with caution. In the other two types, the peat is composed of fibres, these usually being woody. In the coarse variety a mesh of second-order size exists within the interstices of the first-order network, while in fine fibrous peat the interstices are very small and contain colloidal matter.

All present-day surface deposits of peat in Britain have accumulated since the last ice age and therefore have formed during the last 20 000 years. On the other hand, some buried peats may have been developed during inter-glacial periods. Peats also have accumulated in post-glacial lakes and marshes where they are interbedded with silts and muds. Similarly, they may be associated with salt marshes. Fen deposits are thought to have developed in relation to the eustatic changes in sea level, which occurred after the retreat of the last ice sheets. However, blanket bog is a more common type of peat deposit. Valley bogs form along the flatter parts of valley bottoms and generally occur as a result of water draining from relatively acidic rocks.

Fenlands are areas where layers of peat interdigitate with wedges of estuarine silt and clay. The most notable fen deposits in Britain are found south of The

Wash. Some fen peats in Britain, because they occur in areas where chalk or limestone are present, are not associated with acidic water, the acidity being reduced by the lime-rich water contributed to the fen by run-off. Acid fens have poorer plant communities and are less humified because of their acidity than those fen peats where the water is slightly alkaline. In fact, where there is a rich supply of nutrients brought into a peatland by lime-rich surface run-off, the plant communities are much more diverse and give rise to rich fen. Rich fen peat develops a much higher degree of humification than acid peat. Because the strength and permeability of peat declines significantly as humification increases, rich fen peat presents more problems to the engineer than acid fen peat.

Blanket bogs develop on wet uplands where slopes are not excessive and drainage is impeded. The process is often one of paludification and it may start in shallow waterlogged depressions. Bogs extend downslope if poorly drained surface water gives rise to waterlogging. In fact, high rainfall in these areas gives rise to leaching which results in accumulation of impervious humus colloids and iron pan at small distances below the surface, usually between 0.3 and 1.0 m. Such an impermeable layer leads to waterlogging, which represents ideal conditions for the development of peat. Blanket bogs generally are not underlain by soft sands and clays. Generally, these peats are thinner at higher altitudes and on steeper slopes. Significant thicknesses are attained only in large deep depressions in the surface topography. The valley bog has a complex lateral zonation due to the differences in the vegetation developed. For example, it is richer along the borders of the bog and along streams flowing in the valley.

Geological maps show a limited number of peat types and not all of them have accompanying descriptions. The ones currently mapped and included in the British Geological Survey Lexicon of Rock Names are shown in Table 8.

Name	Description
Basin Peat	None
Fen Lower Peat	None
Hill Peat	Peat; of upland areas
Peat	Two main lithologies: 'brushwood' (freshwater) peat and 'phragmites' (brackish water) peat; may be an organic-rich clay; humic deposits. Accumulation of wet, dark brown, partially decomposed vegetation.
Peat Flow	Mass movement of peat resulting from a bog-burst.
-	Other peats are recorded with local names, particularly on more recent maps

Table 8 Types of peat shown on British Geological Survey maps and listed in the BGS Lexicon of Rock Names.

Wastage of peat occurs as a result of permanent drainage works. Drainage lowers the water table, which normally reduces the growth of vegetation and means that oxygen begins to invade the anaerobic zone. This, together with the higher temperature, enhances aerobic decay and associated humification. Furthermore, the effective pressure is raised when the water table declines, causing compression of the peat. For instance, a fall of 1 m imposes an extra load of 10 kPa that can lead to approximately 1.5 m of settlement in a layer of peat 10 m in thickness if the water level is maintained at 1 m below the subsided surface for a year. Capillary suction and desiccation of peat above the water table lead to its shrinkage. As there is no loss of material these processes of reduction in thickness of peat are not included within wastage.

As far as the description of peat is concerned Hobbs (1986) proposed that 10 characteristics should be included in a full description of peat. These were colour, which can provide a guide to the state of the peat (i.e. degree of humification); moisture content (the water content usually increases in amount from fen peat, through transition peat to bog peat); principal plant components, namely, coarse fibre, fine fibre, amorphous granular material and woody material; amount of mineral constituents; smell, notably the distribution of H_2S (methane has no smell); pH value; tensile strength; and any special characteristics.

Depth (m)	Initial void ratio e_o	Coefficient of volume change, m_v (m^2 MN^{-1})		Compression index, C_c		Moisture content (%)	Organic content (%)	Dry density (kN m^{-3})
		A	b	a	b			
1.5	13.38	11.34	2.17	7.02	10.76	894	86.2	1.05
2.0	9.77	8.91	1.46	4.13	5.42	561	67.6	1.55
2.5	9.24	11.12	2.23	4.91	7.87	620	69.0	1.61
3.0	11.61	11.74	2.11	6.38	9.17	795	75.7	1.26
3.5	17.40	11.77	1.94	9.33	12.31	971	61.5	0.94
4.0	10.49	10.26	1.60	5.08	6.34	662	81.6	1.34
4.5	10.26	11.51	1.66	5.58	6.44	583	63.8	1.51
5.5	14.29	10.80	1.57	7.17	8.28	943	79.9	1.02
6.5	16.28	8.76	2.32	6.14	13.82	965	75.6	0.92

Load ranges, σ_v: (a) 12.5-25 kN m^{-2}; (b) 100-200 kN m^{-2}.

Table 9 Some properties of bog peat from Pant Dedwydd, North Wales. (after Nichol and Farmer, 1998).

The organic content of peat varies appreciably, with the mineral content, from over 95% to as low as 50%. For example, Nichol and Farmer (1998) quoted values of 61 to 86% for bog peat from North Wales. As far as engineering is concerned, the organic content is important in that it influences the water-holding capacity of organic soils. The mineral content of organic soils varies, some peat deposits being more or less completely free of mineral matter with dry ash contents as low as 2% to ash contents as high as 50% in some peats found in South Yorkshire (Bell, 1978). The mineral material is usually quartz sand and silt. In many peats the mineral content increases with depth. The specific gravity of peats varies according to the amount of mineral matter contained. The specific gravity of peat has been found to range from as low as 1.1 up to about 1.8, again being influenced by the content of mineral matter.

The void ratio of peat ranges between 9, for dense amorphous granular peat, up to 25, for fibrous types with high contents of sphagnum (Table 9). It usually tends to decrease with depth within a peat deposit. Such high void ratios give rise to phenomenally high water content. The water content of peats varies from a few hundreds per cent dry weight (e.g. 500% in some amorphous granular peats) to over 3 000% in some coarse fibrous varieties. Put another way, the water content may range from 75 to 98% by volume of peat. Moreover, changes in the amount of water content can occur over very small distances. Generally, most water, especially in fibrous peats, occurs as free or intracellular water in the large pores; it also occurs as capillary or interparticle water in the small pores; and as adsorbed osmotic water. Only intracellular and interparticle water can be expelled by consolidation. The proportion of water held in each of these two states, as well as total amount, depends primarily on the morphology and structure of the material present and on the degree of humification. Fen peats have lower interparticle and total water contents than bog peats, mineral constituents reducing the quantity of interparticle water in fen peats.

Gas is formed in peat as plant material decays and this tends to take place from the centre of stems, so gas is held within stems. The volume of gas in peat varies and figures of around 5 to 7.5% have been quoted (Hanrahan, 1954). At this degree of saturation most of the gas is free and so has a significant influence on initial consolidation, rate of consolidation, pore pressure under load and permeability.

The dry density is the important engineering property of peat, influencing its behaviour under load. Hanrahan (1954) recorded dry densities of drained peat within the range 0.63 to 1.26 kN m^{-3}. The dry density is influenced by the mineral content and higher values than that quoted can be obtained when peats possess high mineral residues.

Peat undergoes significant shrinkage on drying out. Nonetheless, the volumetric shrinkage of peat increases up to a maximum and then remains constant, the volume being reduced almost to the point of complete dehydration. The amount of shrinkage that can occur ranges between 10 and 75% of the original volume. The change in peat is permanent in that it cannot recover all the water lost when wet conditions return. Hobbs (1986) noted that the more highly humified peats, even though they have lower water contents, tend to shrink more than the less humified fibrous peat.

The permeability of peat is influenced by the type of plant detritus, degree of humification, and the macro- and micro-structure of peat, although the latter become less significant as the degree of humification increases. It also tends to decline with depth, for example, from 10^{-1} m s^{-1} to 3×10^{-5} m s^{-1} in slightly humified bog peat. Permeability values for highly humified peat may range down to 6×10^{-10} m s^{-1}, although most values fall in the range 1×10^{-5} to 5×10^{-8} m s^{-1}. The permeability of peat is reduced significantly on loading.

Apart from its moisture content and dry density, the shear strength of a peat deposit appears to be influenced by its degree of humification and its mineral content. As both these factors increase so does the shear strength. Conversely, the higher the moisture content of peat, the lower is its shear strength. As the effective weight of 1 m^3 of undrained peat is approximately 45 times that of 1 m^3 drained peat, the reason for the negligible strength of the latter becomes apparent. In an undrained bog the unconfined compressive strength is negligible, the peat possessing a consistency approximating to that of a liquid. The strength is increased by drainage to values between 20 and 30 kPa and the modulus of elasticity to between 100 and 140 kPa. According to Hanrahan (1964) unconfined compressive strengths of up to 70 kPa are not uncommon in peats consolidated under pavements. Due to its extremely low submerged density, which may be between 15 and 35 kg m^{-3}, peat is especially prone to rotational failure or failure by spreading, particularly under the action of horizontal seepage forces.

Consolidation of peat takes place when water is expelled from the pores and the particles undergo some structural rearrangement. Initially, the two processes occur at the same time but as the pore water pressure is reduced to a low value, the expulsion of water and structural rearrangement occur as a creep-like process (Berry and Poskitt, 1972). In other words, the initial stage of drainage can be regarded as primary consolidation whereas the stage of continuing creep represents secondary compression. It must be remembered that primary and secondary consolidation are empirical divisions of a continuous compression process both of which occur simultaneously during part of that process. In fact, accurate prediction of the amount and rate of settlement of peat cannot be derived directly from laboratory tests. Hence large-scale field trials seem to be essential for important projects.

Peat has a high coefficient of secondary compression, the latter being the dominant process in terms of settlement of peat, and in terms of strain, this is virtually independent of water content and degree of saturation. The short phase of primary consolidation is responsible for little distortion. Bog peats appear to possess lower values of secondary compression than fen peats. This probably is because of their non-plastic, highly frictional character.

Differential and excessive settlement is the principal problem confronting the engineer working on a peaty soil. When a load is applied to peat, settlement occurs because of the low lateral resistance offered by the adjacent unloaded peat. Serious shearing stresses are induced even by moderate loads. Worse still, should the loads exceed a given minimum, then settlement may be accompanied by creep, lateral spread, or in extreme cases by rotational slip and upheaval of adjacent ground. At any given time the total settlement in peat due to loading involves settlement with and without volume change. Settlement without volume change is the more serious for it can give rise to the types of failure mentioned. What is more it does not enhance the strength of peat. When peat is compressed the free pore water is expelled under excess hydrostatic pressure. Since the peat is initially quite pervious and the percentage of pore water is high, the magnitude of settlement is large and this period of initial settlement is short (a matter of days in the field).

Because of the potential problem of settlement arising from loading peat, especially in the construction of embankments carrying roads, some method of dealing with peat has to be employed. Bulk excavation of peat frequently is undertaken if the deposit is less than 3 m in thickness. It may be replaced completely or partially by sand and/or gravel, depending on the thickness of the peat. When a deposit exceeds 3 m or peat occurs as layers within soft sediments, precompression commonly is used. The use of precompression involving surcharge loading in the construction of embankments across peatlands involves the removal of the surcharge after a certain period of time. This gives rise to some swelling in the compressed peat. Uplift or rebound can be quite significant depending on the actual settlement and surcharge ratio (i.e. the mass of the surcharge in relation to the weight of the fill once the surcharge has been removed). Rebound also is influenced by the amount of secondary compression induced prior to unloading. The swelling index, C_s, is related to the compression index, C_c, in that an average C_s is around 10% of the C_c, within a range of 5 to 20%. Rebound undergoes a marked increase when surcharge ratios are greater than about 3. Normally rebound in the field is between 2 and 4% of the thickness of the compressed layer of peat before surcharge has been removed. Samson and La Rochelle (1972) showed that the period over which swelling took place was more or less the same as the length of time the surcharge was imposed. With few exceptions improved drainage has no beneficial effect on the rate of consolidation. This is because efficient drainage only accelerates the completion of primary consolidation, which anyhow is completed rapidly.

Deposits of peat have been removed, at times, by blasting. There are three methods of bog blasting. Trench shooting is used where the surface deposits are less than 6 m in depth. One or more rows of charges are placed at the base of the peat and fired to form a trench in which fill is placed. Toe shooting is carried out in peat that is soft and liable to slip. The charges are placed close to fill that has been placed, that is, the peat is removed ahead of the fill. Lastly, in the underfill method, one or more rows of charges are placed beneath fill, the fill then settling into place as the peat is blown aside. Jelisic and Leppanen (1998) referred to the stabilization of peat. The stabilizer is mixed into the peat by a large mixing tool to form columns reminiscent of lime column stabilization in clay. The stabilizing agents may consist of fly ash, ground blast furnace slag or by products from the paper industry.

When peatlands are drained artificially for reclamation purposes the ground level can experience significant subsidence. The subsidence is not simply due to the consolidation, which occurs as a result of the loss of the buoyant force of groundwater but also is attributable to desiccation and shrinkage associated with drying out in the zone of aeration and oxidation of the organic material in that zone. In order to predict subsidence due to drainage of peat deposits, it is necessary to know the thickness, type, density and rate of decomposition of the peat; the position of the water table and the rate and amount of groundwater lowering; the nature of the reclamation; and the type of climate. A classic example in Britain of subsidence of peat due to drainage is provided by the Fenlands where drainage of the peat has taken place over the last 400 years (Waltham, 2000). In some parts of the Fenlands the thickness of peat has been more than halved as a result (e.g. Holme Post was installed in 1848 and by 1988 the thickness of the peat had been reduced from 6.7 m to 2.8 m (Hutchinson, 1980; Waltham, 2000).

Quicksands

As water flows through silts and sands and loses head, its energy is transferred to the particles past which it is moving, which in turn creates a drag effect on the particles. If the drag effect is in the same direction as the force of gravity, then the effective pressure is increased and the soil is stable. Indeed, the soil tends to become more dense. Conversely, if water flows towards the surface, then the drag effect is counter to gravity thereby reducing the effective pressure between particles. If the velocity of upward flow is sufficient it can buoy up the particles so that the effective pressure is reduced to zero. This represents a critical condition where the weight of the submerged soils is balanced by the upward acting seepage force. A critical condition sometimes occurs in silts and sands. If the upward velocity of flow increases beyond the critical hydraulic gradient a quick condition develops in which the soil can undergo a spontaneous loss of strength. This loss of strength causes them to flow like viscous liquids. Terzaghi (1925) explained this phenomenon in the following terms. First, the sand or silt concerned must be saturated and loosely packed. Secondly, on disturbance the constituent grains become more closely packed which leads to an increase in pore water pressure, reducing the forces acting between the grains. This brings about a reduction in strength. If the pore water can escape very rapidly the loss in strength is momentary. Hence, the third condition requires that pore water cannot escape readily. This is fulfilled if the sand or silt has a low permeability and/or the seepage path is long. Casagrande (1936) demonstrated that a critical porosity existed above which a quick condition could be developed. He maintained that many coarse grained sands, even when loosely packed, have porosities approximately equal to the critical conditions, while medium and fine grained sands, especially if uniformly graded, exist well above the critical porosity when loosely packed. Accordingly, fine sands tend to be potentially more unstable than coarse grained varieties. It must also be remembered that the finer sands have lower permeabilities.

Quick conditions brought about by seepage forces are frequently encountered in excavations made in fine sands that are below the water table. As the velocity of the upward seepage force increases further from the critical gradient the soil begins to boil more and more violently. At such a point structures fail by sinking into the quicksand. Liquefaction of potential quicksands also may be brought about by sudden shocks caused by the action of heavy machinery (notably pile driving), blasting and earthquakes. Such shocks increase the stress carried by the water, the neutral stress, and give rise to a decrease in the effective stress and shear strength of the soil. There is also a possibility of a quick condition developing in a layered soil sequence where the individual beds have different permeabilities. Hydraulic conditions are particularly unfavourable where water initially flows through a very permeable horizon with little loss of head, which means that flow takes place under a great hydraulic gradient.

Saturated medium and fine grained sands commonly are found within glacial successions in the north of England, Wales and Scotland. The sands occur as layers or lenses of variable thickness and extent and may be difficult to identify during site investigation. Loose saturated sands are also found in alluvial deposits. Cone penetration testing has been used successfully to identify these sands but may be less successful in glacial deposits because of the possible presence of strong clayey tills or coarse gravel, cobbles and boulders that may impede penetration.

There are several methods that may be employed to avoid the development of quick conditions. One of the most effective techniques is to prolong the length of the seepage path thereby increasing the frictional losses and so reducing the seepage force. This can be accompanied by placing a clay blanket at the base of an excavation where seepage lines converge. If sheet piling is used in excavation of critical soils, then the depth to which it is sunk determines whether or not quick conditions will develop. Consequently, it should be sunk only deep enough to avoid a potential critical condition occurring at the base level of the excavation. The hydrostatic head also can be reduced by means of relief wells and seepage can be intercepted by a wellpoint system placed about the excavation. Furthermore, a quick condition may be prevented by increasing the downward acting force. This may be brought about by laying a load on the surface of the soil where seepage is discharging. Gravel filter beds may be used for this purpose. Suspect soils also can be densified, treated with stabilizing grouts, or frozen. Further details can be found in this volume (Preene, 2001).

Subsurface structures should be designed to be stable with regard to the highest groundwater level that is likely to occur. Structures below groundwater level are acted upon by uplift pressures. If the structure is weak this pressure can cause a blow-out of a basement floor or collapse of a basement wall whereas if the structure is strong but light it may be lifted, that is, subjected to heave. Uplift can be taken care of by adequate drainage or by resisting the upward seepage force. Continuous drainage blankets are effective but should be designed with filters to function without clogging. The entire weight of structure can be mobilized to resist uplift if a raft foundation is used. Anchors, grouted into bedrock, can provide added resistance to uplift.

Glacial deposits

Glacial deposits cover much of lowland Britain and those areas, which were not covered by ice during the Pleistocene period were subjected to periglacial conditions. Two kinds of glacial deposits are distinguished, namely, unstratified drift or till and stratified drift. However, one type commonly grades into the other. Till usually is regarded as being synonymous with boulder clay and is deposited directly by ice while stratified drift is deposited by meltwaters issuing from ice. Generally, these deposits do not present serious problems in geotechnical engineering but at times they can change rapidly and this may present problems, as for example, when clayey tills contain lenses of sand. The nature of a till deposit depends on the lithology of the material from which it was derived, on the position in which it was transported in the glacier, and on the mode of deposition. Till sheets can comprise one or more layers of different material, not all of which are likely to be found at any one locality. Shrinking and reconstituting of an ice sheet can complicate the sequence.

Deposits of till commonly consist of a variable assortment of rock debris ranging from fine rock flour to boulders. They are characteristically unsorted. Some tills may consist essentially of sand and gravel with very little binder. Lenses and pockets of sand, gravel and highly plastic slickensided clay frequently are encountered in tills. Some of the masses of sand and gravel are interconnected, due to the action of meltwater, but many are isolated. The compactness of a till varies according to the degree of consolidation undergone, the amount of cementation and size of the grains. Tills that contain less than 10% clay fraction usually are friable while those with over 10% clay tend to be massive and compact.

Distinction has been made between tills derived from rock debris carried along at the base of a glacier and those deposits, which were transported within or at the terminus of the ice. The former are referred to as lodgement till while the latter is termed ablation till. Lodgement till is commonly stiff, dense and relatively incompressible. Fissures are frequently present in lodgement till, especially if it is clay-matrix dominated. Sub-horizontal fissures have been developed as a result of incremental loading and periodic unloading while sub-vertical fissures owe their formation to the overriding effects of ice and stress relief. Ablation till accumulates on the surface of the ice when englacial debris melts. It therefore is normally consolidated and non-fissile. It is characterized by abundant large stones. The proportion of sand and gravel is high and clay is present only in small amounts (usually less than 10%). The loose packing means that ablation tills often have an low *in situ* density.

Stratified deposits can be subdivided into ice-contact deposits and pro-glacial deposits. Outwash deposits, kames and eskers are examples of the former type of deposits. They usually consist of sands and gravels. The most familiar pro-glacial deposits are varved clays. These sediments accumulated on the floors of glacial lakes and are characteristically composed of alternating laminae of finer and coarser grain size.

Varved clays tend to be normally consolidated or lightly overconsolidated, although it usually is difficult to make the distinction. Bell and Coulthard (1997) found that the average preconsolidation pressure for the Tees Laminated Clay was 253 kPa. This probably is higher than the overburden stress that the clay has undergone throughout its geological history. However, cementation bonding may contribute toward the overconsolidation effect, these laminated clays containing some calcite and dolomite. In addition, their shallow depth means that they probably have been subject to desiccation, as suggested by the fissuring which occurs, especially in the upper layers. This also enhances the overconsolidation effect.

The values of swell index and compression index for the Tees Laminated Clay ranged between 0.11 and 0.68, and 0.24 and 1.55 respectively, indicating a variation from medium to high plasticity (Bell and Coulthard, 1997). If the ratio of the swell index to compression index is less than 0.2, then the soil is non-swelling whereas if it exceeds 0.2, the soils are medium to high swelling. The actual ratios range from 0.25 to 0.45, suggesting that this clay has a potential for swelling, which also was suggested by the activity chart. The varves and laminations of these clays are responsible for their anisotropy (Barden, 1972; and Bell and Coulthard, 1997). The angle of friction varies with the angle of orientation, with a minimum occurring when the laminations are orientated at 45° to the maximum principal stress. Moreover, the anisotropic strength tends to increase with increasing confining pressure. Similarly, the values of the coefficient of consolidation show significant differences in relation to the orientation of the laminations, that is, the coefficient is much larger parallel to the lamination. Although varved/laminated clays can be regarded as more or less impermeable, permeability parallel to the direction of the silty layers can be a magnitude higher.

Frozen soil phenomena

Frozen ground phenomena are found in regions which experience a tundra climate, that is, in those regions where the winter temperatures rarely rise above freezing point and the summer temperatures are only warm enough to cause thawing in the upper metre, or so, of the soil. Although such conditions no longer exist in Britain, they did so during parts of the Pleistocene epoch and immediately after, when ice sheets covered much of the country or were in final retreat, periglacial conditions occurring south of the ice sheets. These conditions gave rise to certain features, which today remain preserved in the ground. Ground may have undergone notable disturbance as a result of mutual interference of growing bodies of ice or from excess pore water pressures developed in confined water-bearing lenses. Involutions are plugs, pockets or tongues of highly disturbed material, generally possessing inferior geotechnical properties, which have been intruded into overlying layers. They are formed as a result of hydrostatic uplift in water trapped under a refreezing surface layer. Frozen soils often display a polygonal pattern of cracks. Individual cracks may be over 1 m wide at their top, may penetrate to depths of 10 m and may be some 12 m apart. They form when, because of exceptionally low temperatures, shrinkage of the ground occurs. Ice wedges occupy these cracks and cause them to expand. When the ice disappears an ice wedge pseudomorph is formed by sediment, frequently sand, filling the crack. Pseudomorphs of ice wedges and involutions usually mean that one material suddenly replaces another. This can cause problems in shallow excavations.

Sheets, lobes and solifluction debris developed under periglacial conditions and transported by mudflow activity, are commonly found at the foot of slopes. These materials may be reactivated by changes in drainage, by stream erosion, by sediment overloading or during construction operations. Solifluction sheets may be underlain by slip surfaces, the residual strength of which controls their stability. Such material in England commonly is referred to as head. Head deposits tend to be relatively permeable, weak and compressible materials. However, their strength is affected by the proportion of rock fragments they contain. Those deposits derived from calcareous rocks may be weakly cemented by calcium carbonate.

Frost action in a soil, of course, is not restricted to tundra regions. Its occurrence is influenced by the initial temperature of the soil, as well as the air temperature, the intensity and duration of the freeze period, the depth of frost penetration, the depth of the water table, and the type of ground and exposure cover. Settlement is associated with thawing of frozen ground. As ice melts, settlement occurs, water being squeezed from the ground by overburden pressure or by any applied loads. Excess pore pressures develop when the rate of ice melt is greater than the discharge capacity of the soil, and can lead to the failure of slopes and foundations.

Capillary saturation at the beginning and during the freezing of the soil, a plentiful supply of subsoil water, and a soil possessing fairly high capillarity together with moderate permeability are necessary for the occurrence of frost heave. Furthermore, the ground surface experiences an increasing larger amount of heave, the higher the initial water table. Grain size is another important factor influencing frost heave. For example, gravels, sands and clays are not particularly susceptible to heave while silts definitely are. The reason for this is that silty soils are associated with high capillary rises but at the same time their voids are large enough to allow moisture to move quickly enough for them to become saturated rapidly. Casagrande (1932) suggested that the particle size critical to heave formation was 0.02 mm. If the quantity of such particles in a soil is less than 1%, no heave is to be expected, but considerable heaving may take place if this amount is over 3% in non-uniform soils and over 10% in very uniform soils.

Croney and Jacobs (1967) suggested that under the climatic conditions experienced in Britain well-drained cohesive soils with a plasticity index exceeding 15% could be looked upon as non-frost susceptible. They suggested that where the drainage is poor and the water table is within 0.6 m of formation level the limiting value of plasticity index should be increased to 20%. In addition, in experiments with sand, they noted that as the amount of silt added was increased up to 55% or the clay fraction up to 33%, the decrease in permeability in the freezing front was the overriding factor and heave tended to increase. Beyond these values the decreasing permeability below the freezing zone became dominant and progressively reduced the heave. This indicates that the permeability below the frozen zone was principally responsible for controlling heave.

Where there is a likelihood of frost heave occurring it is necessary to estimate the depth of frost penetration. Once this has been done, provision can be made for the installation of adequate insulation or drainage within the soil and to determine the amount by which the water table may need to be lowered so that it is not affected by frost penetration. The base of footings should be placed below the estimated depth of frost penetration, as should water supply lines and other services. Frost susceptible soils may be replaced by gravels. The addition of certain chemicals to soil can reduce its capacity for water absorption and so can influence frost susceptibility. For example, Croney and Jacobs (1967) noted that the addition of calcium lignosulphate and sodium tripolyphosphate to silty soils were both effective in reducing frost heave.

The influence of weathering and fissures on clay soils

The greatest variation in the engineering properties of clays can be attributed to the degree of weathering which they have undergone. Generally, changes in the clay mineral and quartz content of clay deposits on weathering are slight. However, the degree of illite degradation increases in weathered clay and that with the most degraded illite appears to have the lowest strength. In addition, consolidation of a clay deposit gives rise to an anisotropic texture due to the rotation of the platey minerals and diagenetic processes bond particles together. Weathering reverses these processes, altering the anisotropic structure and destroying or weakening interparticle bonds. Therefore, weathering, through the destruction of interparticle bonds, leads to a clay deposit reverting to a normally consolidated sensibly remoulded condition. Higher moisture contents are found in the more weathered clay. This progressive degrading and softening also is accompanied by reductions in strength and deformation moduli with a general increase in plasticity. For example, Cripps and Taylor (1981) indicated that the undrained shear strength of the London Clay is reduced by approximately half on weathering. The removal of overburden leads to vertical expansion of a deposit, which facilitates the development of joints and fissures, together with softening. The opening of fissures is accompanied by water entrainment and chemical degradation.

Chandler (1972) recognized four zones of weathering in the Upper Lias Clay. A similar situation was found in the Lower Lias Clay by Coulthard and Bell (1993) who recorded several notable changes in the composition and basic geotechnical properties of the clay as the degree of weathering increased. Changes also were noted in the fabric and mesostructures. The strength-depth relationship in the Lower Lias Clay was not related to variation in moisture content but was attributable to the decreasing number of fissures with depth. Russell and Parker (1979) recognized a similar zoning of the grades of weathering in the Oxford Clay. They found a reasonable correlation between shear strength and moisture content, although this was not as good as in the Upper Lias Clay, probably because of the greater variability of the Oxford Clay.

Fissures play an extremely important role in the failure mechanism of fissured clay. Indeed, many clays are seriously weakened by the presence of a network of fissures. Terzaghi (1936) provided the first quantitative data relating to the influence of fissures and joints on the strength of clays, pointing out that they are characteristic of overconsolidated clays. He maintained that fissures in normally consolidated clays have no significant practical consequences. On the other hand, fissures can have a decisive influence on the engineering performance of an overconsolidated clay, in that the overall strength of such fissured clay can be as low as one-tenth of that of the intact clay. In addition to allowing clay to soften, fissures and joints allow concentrations of shear stress, which locally exceed the peak strength of clay, thereby giving rise to progressive failure. Under stress, the fissures in clay seem to propagate and coalesce in a complex manner.

Skempton and La Rochelle (1965) investigated slope failures in the London Clay at Bradwell, Essex. These began to occur on cut slopes a few days after excavation, even though the slopes had been designed on strengths lower than the undrained parameters for the intact clay. The investigation revealed that a large proportion of the reduction in the overall strength below these design parameters was attributed to the fissures. In fact, the strength along joints or fissures in clay is normally only slightly higher than the residual strength of the intact clay. Skempton and La Rochelle considered that where unfavourable orientation of fissures exists, the major part of a failure plane may follow the fissures. As a result, the clay may be near its residual strength. The overall strength of the clay could be further reduced by the separation of the walls of closed fissures. As intact clay has a low tensile strength, there is no resistance to the opening of fissures and once open there is no shear resistance along them. They recommended that, if regular fissure patterns with an unfavourable orientation occur at a site, an attempt should be made to estimate the influence of these fissures on the overall strength of the clay mass. Therefore, it is necessary to determine the average area of a potential failure plane, which would pass through open fissures, closed fissures and intact clay.

Conclusions

The degree to which a soil is problematic to engineering is a function of the nature of the soil itself (mineralogy, micro-fabric and geotechnical and other properties), the geological processes that caused it to be formed (whether fluvial, glacial, aeolian, or others) and the processes that are acting on it currently (weathering, erosion and human activity). Most of these problematic soils are young in geological terms. Collapsible soils, peat soils, quicksands, glacial deposits, frozen soil phenomena are all Quaternary in age; in Britain, only expansive clays are (mostly) older. Since their formation, weathering and erosive processes and have altered some of these soils. For example, overconsolidated clays become fissured and lose strength as a result of weathering. Therefore, in assessing the risk that these soils pose for engineering structures, all these factors must be taken into account.

The main cause of expansiveness in clay soils is the presence of swelling clay minerals such as montmorillonite. The potential for volume change is controlled by the initial moisture content, initial dry density or void ratio, the microstructure and the vertical stress, as well as the type and quantity of clay minerals present. In Britain, the most expansive clay soils are of Mesozoic age and younger and are found mainly in the south, south east and Midlands of England. Methods for predicting volume change can be grouped into empirical ones (principally assessing plasticity), soil suction methods and oedometer testing. Three main methods have been adopted to mitigate the effects of swelling and shrinkage: use of foundations and structures that can tolerate movements without unacceptable damage; isolation of the foundation and structure from the effects of the soil; control of ground conditions, particularly moisture content.

Certain soils, composed mainly of silt and usually wind-blown in origin, have the potential to collapse, that is, they are susceptible to large reductions in void ratio when wetted, either under overburden pressure or additional imposed load (for example, by a building). These soils have porous textures and high void ratios and relatively low densities. The fabric of collapsible soils generally consists of a loose skeleton of grains (usually quartz) and/or aggregations of clay and silt. The particles or aggregations tend to be separate from each other but connected by bonds and bridges of clay-sized minerals. The main collapsible soil in Britain is brickearth, which is found principally in south and south eastern England. A wide range of collapse criteria have been published based, usually, on various combinations of void ratio, dry density and plasticity. Alternatively, collapsibility can be determined in the oedometer by loading a sample at natural moisture content to a given load and then flooding the sample and measuring the collapse strain. Methods for treating collapsible soils include moistening and compaction, over-excavation and recompaction, vibro-flotation, vibroreplacement, dynamic compaction, lime injection or piles, jet grouting, ponding or flooding and heat treatment.

Peat consists of partially decomposed and disintegrated plant remains that have been preserved in conditions of incomplete aeration and high water content. It is found where there is an excess of rainfall and poorly drained ground. Consequently, it is found in many parts of Britain, both lowland and upland. Peat usually has an organic content well in excess of 50%, void ratios between 9 and 35 and water contents over 500%. The main engineering problems associated with peat are differential and excessive settlement. During construction the principal methods of treatment are bulk excavation if the peat is less than about 3 m thick, the application of surcharge loads to embankments with subsequent removal after a period of time, removal by blasting and stabilization using fly ash, furnace slag or paper industry by-products.

Saturated sands and silts become 'quick' (and flow) when the weight of the submerged soil is balanced by the upward acting seepage force and so the effective pressure is reduced to zero. Finer sands and coarser silts tend to be the more susceptible materials when saturated and loosely packed because they have lower permeabilities than coarser sands and so excess pore pressures are less easily dissipated. Liquefaction can be brought about by sudden shocks caused by heavy machinery (such as pile drivers), blasting and earthquakes. Flow can also take place in layered soil sequences where individual beds have different permeabilities. Susceptible soils are found within glacial sequences in northern Britain and in alluvial deposits. Quick conditions can be treated by extending the length of the flow path and so increasing frictional losses. Hydrostatic heads can be reduced by relief wells and seepage intercepted by a wellpoint system around an excavation. Densification, use of stabilization grouts and ground freezing are other methods of treatment.

Glacial and periglacial deposits pose a number of problems. While many glacial tills are strong and provide good foundations, they may contain laminated clays formed in glacial lakes. Unlike the tills, these are usually normally consolidated or lightly overconsolidated and may have a tendency to swell. They are likely to show greater permeability horizontally. Solifluction deposits formed in a periglacial environment may be underlain by slip surfaces, the residual strength of which controls their stability. Silts may be susceptible to frost heave because the soils possess fairly high capillarity together with moderate permeability, allowing moisture to move quickly enough for them to become saturated rapidly. Footings should be placed below the depth of frost penetration in susceptible soils or the soils can be chemically treated.

Weathering of clay soils leads to a loss of strength, reduction of deformation moduli and increases in plasticity. Fissuring also increases, which reduces mass strength.

Acknowledgements

This paper is published with the permission of the Director of the British Geological Survey (NERC).

References

Anon. 1980. *Low-rise Buildings on Shrinkable Clay Soils, Part 3.* Building Research Establishment, Digest 240, Her Majesty's Stationery Office, London, 8p.

Barden, L. 1972. *The influence of structure on deformation and failure of clay soil.* Geotechnique, 22, 159-163.

Bell, F.G. 1978. *Peat: a note on its geotechnical properties.* Civil Engineering, January 45-49 and February, 49-53.

Bell, F.G. and Coulthard, J.M. 1997. *A survey of some geotechnical properties of the Tees Laminated Clay of central Middlesbrough, north east England.* Engineering Geology, 48, 117-133.

Bell, F.G. and Maud, R.R. 1995. *Expansive clays and construction, especially of low rise structures: a view point from Natal, South Africa.* Environmental and Engineering Geoscience, 1, 41-59.

Berry, P.L. and Poskitt, T.J. 1972. *The consolidation of peat.* Geotechnique, 22, 27-52.

Burland, J.B. and Wroth, C.P. 1975. *Allowable and differential settlement of structures including damage and soil structure interaction.* In Settlement of Structures, British Geotechnical Society, Pentech Press, London, 611-654.

Casagrande, A. 1932. *Discussion on frost heaving.* Proceedings Highway Research Board, Bulletin No. 12, 169.

Casagrande, A. 1936. *Characteristics of cohesionless soils affecting the stability of slopes and earth fills.* Journal Boston Society Civil Engineers, 23, 3-32.

Chandler, R.J. 1972. *Lias Clay: weathering processes and their effect on shear strength.* Geotechnique, 22, 403-431.

Chandler, R.J., Crilly, M.S. and Montgomery-Smith, G. 1992. *A low cost method of assessing clay desiccation for low-rise buildings.* Proceedings Institution Civil Engineers, 92, 82-89.

Chen, F.H. 1988. *Foundations on Expansive Soils.* Elsevier, Amsterdam.

Clemence, S.P and Finbarr, A.O. 1981. *Design considerations for collapsible soils.* Proceedings American Society Civil Engineers, Journal Geotechnical Engineering Division, 107, 305-317.

Clevenger, W.A. 1958. *Experience with loess as foundation material.* Proceedings American Society Civil Engineers, Journal Soil Mechanics Foundations Division, 85, 151-180.

Coulthard, J.M. and Bell, F.G. 1993. *The engineering geology of the Lower Lias Clay at Blockley, Gloucestershire, U.K.* Geotechnical and Geological Engineering, 11, 185-201.

Cripps, J.C. and Taylor, R.K. 1981. *The engineering properties of mudrocks.* Quarterly Journal Engineering Geology, 14, 325-346.

Croney, D. and Jacobs, J.C. 1967. *The Frost Susceptibility of Soils and Road Materials.* Transport Road Research Laboratory, Report LR90, Crowthorne.

Denisov, H.Y. 1963. *About the nature and sensitivity of quick clays.* Osnov Fudamic Mekhanic Grant, 5, 5-8

Derbyshire, E. and Mellors, T.W. 1988. *Geological and geotechnical characteristics of some loess and loessic soils from China and Britain: a comparison.* Engineering Geology, 25, 135-175.

Driscoll, R. 1983. *The influence of vegetation on the swelling and shrinkage of clay soils in Britain.* Geotechnique, 33, 93-105.

Evstatiev, D. 1988. *Loess improvement methods.* Engineering Geology, 25, 341-366.

Feda, J. 1966. *Structural stability of subsidence loess from Praha-Dejvice.* Engineering Geology, 1, 201-219.

Fookes, P.G. and Best, R. 1969. *Consolidation characteristics of some late Pleistocene periglacial metastable soils of east Kent.* Quarterly Journal Engineering Geology, 2, 103-128.

Gibbard, P.L. 2001. *Per. Comm.*

Gibbs, H.H. and Bara, J.P. 1962. *Predicting surface subsidence from basic soil tests.* American Society Testing Materials (ASTM), Special Technical Publication, No 322, 231-246.

Gibbs, H.H. and Bara, J.P. 1967. *Stability problems of collapsing soil.* Proceedings American Society Civil Engineering, Journal Soil Mechanics Foundations Division, 93, 572-594.

Gourley, C.S., Newill, D. and Schreiner, H.D. 1994. *Expansive soils: TRL's research strategy.* In Engineering Characteristics of Arid Soils, (Fookes, P.G. and Parry, R.H.G., eds.), A.A. Balkema, Rotterdam, 247-260.

Grabowska-Olszewsla, B. 1988. *Engineering geological problems of loess in Poland.* Engineering Geology, 25, 177-199.

Grim, R.E. 1962. *Applied Clay Mineralogy,* McGraw-Hill, New York.

Handy, R.L. 1973. *Collapsible loess in Iowa.* Proceedings American Society Soil Science, 37, 281-284.

Hanrahan, E.T. 1954. *An investigation of some physical properties of peat.* Geotechnique, 4, 108-123.

Hanrahan, F.T. 1964. *A road failure on peat.* Geotechnique, 14, 185-203.

Hobbs, N.B. 1986. *Mire morphology and the properties and behaviour of some British and foreign peats.* Quarterly Journal Engineering Geology, 19, 7-80.

Hutchinson, J.H. 1980. *The record of peat wastage in the East Anglian fenlands at Holme Post, 1848-1978AD.* Journal of Ecology, 68, 229-249.

Jelisic, N. and Leppanen, M. 1998. *Mass stabilization of peat in road and railway construction.* Proceedings 8th International Congress International Association Engineering Geology, Vancouver, A.A. Balkema, Rotterdam, 5, 3449-3454.

Jennings, J.E. and Knight, K. 1957. *The prediction of total heave from the double oedometer test.* Transactions South African Institution Civil Engineers, 7, 285-291.

Jennings, J.E. and Knight, K. 1975. *A guide to construction on or with materials exhibiting additional settlement due to collapse of grain structure.* Proceedings Sixth African Conference Soil Mechanics Foundation Engineering, Durban, 99-105.

Landva, A.O. and Pheeney, P.E. 1980. *Peat fabric and structure.* Canadian Geotechnical Journal, 17, 416-435.

Lin, Z.G. and Wang, S.J. 1988. *Collapsibility and deformation characteristics of deep-seated loess in China.* Engineering Geology, 25, 271-282.

Litvinov, I.M. 1973. *Deep compaction of soils with the aim of considerably increasing their bearing capacity.* Proceedings Eighth International Conference Soil Mechanics and Foundation Engineering, Moscow, **3**, 271-282.

Lutenegger, A.J. and Hallberg, G.R. 1988. *Stability of loess.* Engineering Geology, 25, 247-261.

McQueen, I.S. and Miller, R.F. 1968. *Calibration and evaluation of wide ring gravimetric methods for measuring moisture stress.* Soil Science, 106, 225-231.

Nichol, D. and Farmer, I.W. 1998. *Settlement over peat on the A5 at Pant Dedwydd near Cerrigydrundion, North Wales.* Engineering Geology, 50, 299-307.

Northmore, K.J., Bell, F.G. and Culshaw, M.G. 1996. *The engineering properties and behaviour of the brickearth of south Essex.* Quarterly Journal Engineering Geology, 29, 147-161.

O'Neill, M.W. and Poormoayed, A.M. 1980. *Methodology for foundations on expansive clays.* Proceedings American Society Civil Engineers, Journal Geotchnmical Engineering Division, 106, 1345-1367.

Phien-Wej, N., Pientong, T. and Balasubramanian, A.S. 1992. *Collapse and strength characteristics of loess in Thailand.* Engineering Geology, 32, 59-72.

Popescu, M.E. 1986. *A comparison of the behaviour of swelling and collapsing soils.* Engineering Geology, 23, 145-163.

Preene, M. 2001. *Ground improvement by dewatering.* In Problematic Soils Symp., (Jefferson, I., Murray, E.J., Faragher, E., and Fleming, P.R., eds.), Nottingham (*in press)*

Radforth, N.W. 1952. *Suggested classifications of muskeg for the engineer.* Engineering Journal (Canada), 35, 1194-1210.

Russell, D.J. and Parker, A. 1979. *Geotechnical, mineralogical and chemical inter-relationships in weathering profiles of an overconsolidated clay.* Quarterly Journal Engineering Geology, 12, 197-216.

Samson, L. and La Rochelle, P. 1972. *Design and performance of an expressway constructed over peat by preloading.* Canadian Geotechnical Journal, 9, 447-446.

Skempton, A.W. and La Rochelle, P. 1965. *The Bradwell slip: a short term failure in London Clay.* Geotechnique, 15, 221-241.

Terzaghi, K. 1925. *Erdbaumechnik auf Boden Physikalischer Grundlage.* Deuticke, Vienna. (in German)

Terzaghi, K. 1936. *Stability of slopes of natural clay.* Proceedings 1st International Conference Soil Mechanics Foundation Engineering, Cambridge, Mass., 1, 161-165.

Van Der Merwe, D.H., 1964, *The prediction of heave from the plasticity index and the percentage clay fraction*, The Civil Engineer in South Africa, 6, No. 6, 103-107.

Williams, A.A.B. and Pidgeon, J.T. 1983. *Evapotranspiration and heaving clays in South Africa.* Geotechnique, 33, 141-150.

Waltham, A.C. 2000. *Peat subsidence at the Holme Post.* Mercian Geologist, 15 (1), 49-51

Behaviour of silt: the engineering characteristics of loess in the UK

I. Jefferson[1] , **C. Tye**[1] **and K.J. Northmore**[2]
1) *School of Property and Construction, Nottingham Trent University, Nottingham, NG1 4BU*
2) *Urban Geoscience and Geological Hazards Programme, British Geological Survey, Nottingham, NG12 5GG*

Introduction

Loess soils cover a large area (approximately 10%) of the Earth's land-mass. Loess consists essentially of silt-sized (typically 20-30 microns) primary quartz particles that form as a result of high energy earth-surface processes such as glacial grinding or cold climate weathering (Rogers *et al.,* 1994). These particles are transported from the source (such as tectonically active mountains, e.g. Himalayan and Alpine ranges) by great rivers (e.g. Hwang He, the Danube and the Rhine). Subsequent flooding of these rivers allows the quartz silt particles to be deposited on flood plains. On drying out these particles are detached and transported by the prevailing winds until deposition leeward at distances ranging from tens to thousands of kilometres. This process has resulted in the almost continuous deposit draped from the North China plain to southeast England (where it is generally referred to as 'brickearth'). Although it is possible to isolate five major regions: North America, South America, Europe including western Russia, Central Asia and China. These loess regions underlie highly populated areas and major infrastructure links, and are structurally metastable, that is, the deposits are prone to rapid collapse settlement leading to ground subsidence. The areas of most widespread concern are concentrated in Eastern Europe and Russia and to a growing extent in China (see Derbyshire *et al.*, 1995), although serious problems of potential collapse exist wherever loess is found.

The particles, which make up loess deposits, although principally of silt-sized quartz, consist also of feldspars and micas. Clay-sized particles within the loess structure consist of quartz, feldspar and carbonates in addition to true clay minerals. This compositional picture is complicated further by differences (particularly mineralogical) in the clay-sized fraction in loess and palaeosols (old buried soils), both between different climatic regions and between loess

Problematic Soils. Thomas Telford, London, 2001

units and buried palaeosol horizons within the same climate environment. These differences are a fundamental cause of variation in metastability, and hence the potential collapse that may result after a loess soil is loaded and/or wetted. In addition, the primary quartz particles are irregular in shape (Rogers and Smalley, 1993). As a result of their genesis and constitution, loess deposits form remarkably open structures with the interstitial clay-sized particles congregating at the quartz particle contacts. A process of inter-particle bonding, the strength of which can in certain circumstances increase with time, maintains this open structure.

Loess can cause a number of engineering problems. These problems result because loess undergoes structural collapse and subsidence due to saturation when both the initial dry density and initial water content are low (Rogers *et al.,* 1994). Generally, soils that possess porous textures with high void ratios and relatively low densities have the potential to collapse. They have sufficient void space in their natural state to hold their liquid limit moisture content when saturated. At their natural moisture content these soils possess high apparent strength but they are susceptible to large reductions in void ratio on wetting, especially under load. In other words, the metastable texture collapses as the bonds between the grains break down when the soil is wetted. Hence, the collapse process represents a rearrangement of soil particles into a denser state of packing. Collapse on saturation usually takes only a short period of time. As such, the soil passes from an under-consolidated condition to one of normal consolidation. Such soils frequently are of aeolian origin, and of silt size with uniform grading. Both loess and brickearth have glacial associations in that it is believed that these silty soils were derived from continental areas where silty material was produced by glacial action prior to aeolian transportation and deposition.

This paper will give a brief overview of the occurrence, properties and engineering behaviour of loess soils in the UK. This will include discussion on its formation and the use of the term brickearth and the problems this nomenclature has generated. This paper complements the discussion presented by Bell and Culshaw (2001) earlier in this volume.

Distribution and occurrence of loess in the UK

Figure 1 indicates the approximate distribution of loess/brickearth greater than 0.3m in thickness in the UK. Significant thicknesses (greater than 1m) are restricted to north and east Kent (see Fookes and Best, 1969, Derbyshire and Mellors, 1988, and Dibben *et al.,* 2001), south Essex (see Northmore *et al.,* 1996, Miller *at al.,* 2001) and the Sussex coastal plains. In Essex, deposits of up to 8m have been found, although thicknesses of 4m or so are more typical (Northmore *et al.,* 1996). Regional trends in the type of deposit can be determined, due to textural and mineralogical distinctions. However, a progressive westward increase in the proportion of heavy flaky minerals (chlorite and biotite) is typically found. Moreover, there is a progressive westward decrease in modal size, which has been suggested as being due to sorting by easterly winds during deposition, from source material in the North Sea Basin or near continent (Catt, 1985). Originally, loessic deposits would have been more extensive but have since been removed by post depositional erosion, colluviation, deforestation/agriculture and resource stripping activities (Catt, 1977).

Figure 1 Distribution of 'brickearth' deposits in southern England and Wales (modified after Catt, 1988).

In the United Kingdom the deposit mapped as brickearth is similar to loess and occurs as discontinuous 'spreads' in southern and eastern England, notably in Essex, Kent, Sussex and Hampshire (Figure 1). In addition, similar material occurs in widened fissures (gulls) in deposits such as the Hythe Beds in Kent and in sinkholes in the Chalk north-west of London (Bell *et al.*, 2001). Whether such fill is of windblown origin or whether it is derived locally from existing silty deposits that have undergone solifluction is still a matter of debate.

Many small pockets of loess are found in horizons up to 1m in thickness. These include loess-derived material on the Carbonifererous limestones in Somerset and South Wales and on clay-with flint and related superficial deposits covering the chalk plateau of southern England and Tertiary deposits in Hampshire. Other areas of loessic deposits occur around Heathrow (see Rose *et al.*, 2000), on the Devonian Limestone and Permian Breccias in Devon (Cattell, 1997) and in many coastal areas, e.g. Cornwall (see Catt and Stains, 1982, and Roberts, 1985) and the Wirral (see Lee, 1979). In Britain, almost all the brickearth/loess soils are probably of late Devensian age (c. 14 000 to 30 000 years BP) (Kerney, 1965).

Loess and brickearth

In the United Kingdom, a nomenclature problem has arisen with the use of the term *brickearth*. Literally it means the earth used for manufacture bricks, however not all earth used in brick manufacture is brickearth and not all brickearth is used for the manufacture of bricks. An example of this is the use of the term to describe river alluvium of Pleistocene age (Gibbons, 1981). Historically, this term has been used to describe a particular type of Quaternary deposit that has distinctive lithological characteristics and engineering properties. Additional confusion has been introduced by the use of geological descriptors on the survey sheets for the brickearth-like deposits. These other deposits have very similar lithological and engineering characteristics to those simply designated as 'brickearth' but differ in their topographical position. The variation in terminology is intended to reflect the various fluviatile and aeolian (wind-blown) environments in which these deposits were laid down, for example Head Brickearth, River Loam/ River Terrace Loam and River Brickearth (see Bell and Culshaw, 2001). Subsequent remobilisation by cryoturbation and solifluction together with sub-aqueous redistribution have led to the evolution of complex modified sequences (Northmore and Hobbs, 1999).

Formation and material characteristics

Wind-blown deposits of loess are typically characterised by a lack of stratification and uniform sorting, and occur as blanket deposits. Loess may exhibit sub-vertical columnar jointing and in addition, pipe systems may be developed. Because deposits of loess show a close resemblance to fine-grained glacial debris, their origin has generally been assigned a glacial association. In essence, winds blowing from the arid interiors of the northern continents during glacial times picked up fine glacial outwash material and carried it many hundreds of kilometres before deposition took place. Deposition is assumed to have occurred over steppe lands, the grasses having left behind fossil root-holes, which typify loess (Pye, 1987). The lengthy transport accounts for the uniform sorting of loess, with generally greater sorting occurring in a westward direction in UK Loess soils. It should be noted that other alluvial and periglacial process in the formation of loess can also occur which result in modification of the original wind blown deposit.

In spite of having a variety of origins and appreciable differences in provenance, thickness and age, loess is a remarkably uniform soil in terms of its dominant minerals and geotechnical behaviour. In other words, loess deposits frequently have similar grain size distribution, mineral composition, open texture, low degree of saturation and bonding of grains that is not resistant to water.

The fabric of loess takes the form of a loose skeleton built of grains (generally quartz) and microaggregations (assemblages of clay or clay and silty clay particles). The sand and silt-sized particles are sub-angular to sub-rounded, and may be in grain to grain contact or separate from each other, being connected by bonds and bridges (see Figure 2). The bridges are formed of clay-sized materials, (clay minerals, fine quartz, feldspar or calcite). These clay-sized materials also occur as coatings to grains. Silica and iron oxide may be concentrated as cement at grain contacts, and amorphous overgrowths of silica occur on grains of quartz and feldspar. The nature and quality of the bonds connecting the silt particles govern the mechanical behaviour of loess.

Figure 2 Scanning Electron Micrograph of bonded quartz particles in Brickearth (loess) from Ockley brickworks, Sittingbourne, Kent.

In the southern UK, the deposits of loess/brickearth are found on landscapes beyond the limits of where the glaciers reached in Devensian times. Winds blowing off the continental ice masses during the glaciations of the Pleistocene and over outwash plains, hill deposits and frost shattered surfaces, picked up material and carried it in suspension away from the ice. The majority of the deposits in southern UK are generally thought to represent the 'tail-end' of the European loess belt. Lill and Smalley (1978) argued that easterly winds driven by large anticyclones over the Scandinavian ice sheets may have carried silt to England, and may explain the relative abundance loess in southeastern England. Eden's (1980) investigations of Norfolk, Essex and Kent loess suggested a North Sea Basin source with winds moving from the north. The loessic deposits in the south-west of the UK are thought to be derived from glacigenic sediment in the southern Irish

Sea Basin that also acted as the source area for deposits found in north/west England (Catt and Staines, 1982). Relief is seen to have an influence on the loessic deposits in Essex, as the deposit tends to mantle gentle slopes with easterly or northerly aspects. Evidence of 'river brickearth'/ 'loam' is observed on the flat tops of terraces, but if found above terrace level the material is called brickearth. Therefore, this arbitrary distinction may have led to loessic deposits being classed as 'terrace loams'.

Following aeolian transportation, deposition of the silt particles occurs by settling through an air column, trapping by vegetation, rough or wet surface and by rainfall events. These processes are thought to give rise to an 'open' soil fabric, however any fabric is influenced by post-depositional history of the soil. The most important post-depositional influence is that of cyclic wetting and drying produced by seasonal soil moisture fluctuations which leads to densification, secondary (authigenic) clay development and illuviation and cementation. Moisture loss on drying (desiccation) is equivalent to a loading process as it increases the intergranular stresses through suction. This can produce a more densely packed structure (Derbyshire and Mellor, 1986). Therefore, estimating the mechanisms that have fashioned the distribution may also help to explain the subsequent engineering characteristics.

Figure 2 shows silt-size particles coated and bridged by platy clay particles. Brickearth generally is calcareous at depth, although upper layers may often be leached of carbonate material. The lower parts of a deposit tend to be better consolidated and more rigid than the upper parts. Quartz is the most abundant mineral in brickearth followed by feldspar (Bell *et al.* 2001). Of the clay-type minerals, mica is generally the most abundant, then montmorillonite followed by illite and kaolinite. Calcium carbonate occurs as grains, thin tube infillings and as concretionary nodules. When present it tends to account for less than 10% of the soil. However, in Kent, values as high as 20% have been found (Derbyshire and Mellors, 1988).

The principal mineral (up to 90%) in some gull-fills found in Kent is quartz. The quartz grains are sub-angular and sub-rounded to rounded, with the degree of roundness tending to increase with increasing size of grains. The remainder of the fill usually consists of sub-angular flint grains, hornblende, some glauconite and traces of heavy minerals and micaceous material. The carbonate content varies between 0.03% and 0.46%, which is very low compared with brickearth from south Essex and Kent. The small amount of calcium carbonate may be due to the material having been leached.

Properties of loess in UK

The loess/brickearth is frequently described as a pale yellowish-brown or orange-brown slightly plastic, unbedded (non- or poorly stratified), highly porous, variably clayey SILT, often displaying well developed vertical prismatic jointing. It is generally calcareous, often with a well-defined, leached non-calcareous upper zone overlying a lower calcareous layer. Small calcareous tubular concretions, often containing rootlet traces, are common (Northmore and Hobbs, 1999). Using the

BGS's 'Quaternary Province Classification System' these deposits are found in lowland periglaciated areas (Foster *et al.,* 1999).

Loess/brickearth deposits generally consist of 50 to 90% silt sized quartz particles. The occurrence of calcite and clay is periodic throughout a deposit. Brickearth is generally well graded with a consistently fine skew (Figure 3). Both loess and brickearth can be classified as clayey, silty or sandy, depending on the fine content present. Although clay contents vary throughout the UK, silt is still the primary constituent. The particle size distribution of gull-fill material from Allington, Kent indicates that it falls within the clayey loess and silty loess zones. This means that it has affinities with brickearth from Pegwell Bay, Kent, as reviewed by Fookes and Best (1969), as well as with those from other locations in Kent and in south Essex.

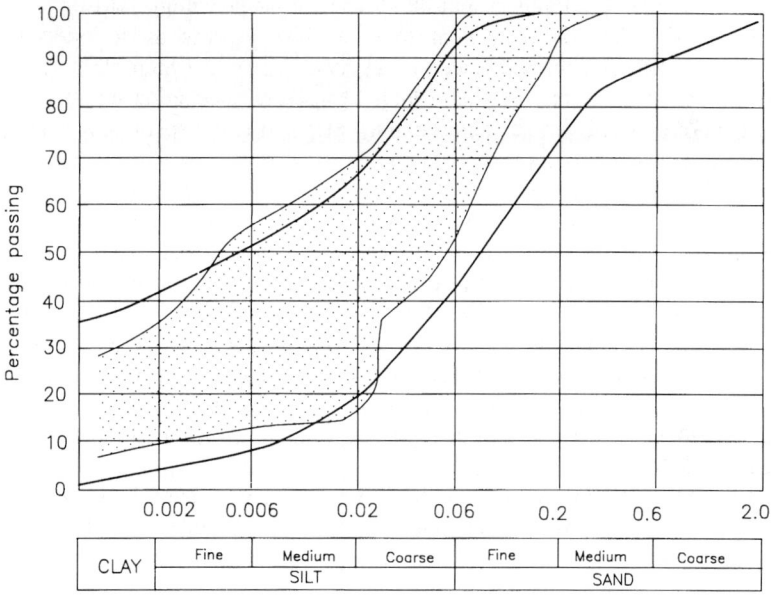

Figure 3 Particle size distribution of brickearth from south Essex compared with that from Kent (shown stippled), (after Northmore *et al.* 1996).

It is the arrangement of particles (i.e. the fabric) which imparts the characteristic engineering properties, particularly that of metastability to these loessic soils. In the UK, the material is highly porous (typically with void ratios of 0.5 – 0.9) and so the deposits are free-draining and tend to remain partially saturated even through the wet British winter. If maintained, these deposits have a relatively high strength and hence natural stability in vertical sections or excavations, see Figure 4.

Figure 4 An example of loess in the UK – Pegwell Bay, Kent. The cliff exposure is 3m of loess overlying chalk

Other soil and index properties have been reviewed elsewhere (see Fookes and Best, 1969; Derbyshire and Mellors, 1988, Boardman *et al.,* 2001; Bell and Culshaw, 2001).

Strength, consolidation and permeability of loess and brickearth

The strength of loess/brickearth is dependent on the initial porosity and moisture content, the degree of deterioration of the bonds and the increase in granular contacts under consolidation, as well as changes in moisture content. When loess/brickearth with many macropores and high water content is loaded, the 'cementing' bonds are first broken, resulting in a lowering of the cohesion and a softening of the soil. With further loading, the grains are brought more and more into contact, thereby increasing friction, so giving rise to a hardening effect.

Bell and Culshaw (2001) discuss results presented by Northmore *et al.,* (1996) for undrained shear strength (measured in the triaxial compression test). A significant degree of variability was observed in the results presented and this was considered to be related partly to variability of composition. Near-surface strengths are higher, reflecting the desiccated, 'crust-like' nature of the loess at the ground surface. However, Northmore *et al.* noted a general tendency for shear strength to decrease with increasing depth.

Loess/brickearth can support heavy structures with small settlements if loads do not exceed the apparent preconsolidation stress and natural moisture content is low. On the other hand, loess can compress substantially if the apparent preconsolidation stress is exceeded. Bell and Culshaw (2001) provide details of the compressibility of loess/brickearth in the UK, with compression indices of between 0.33 to 0.15 being observed (cf. Table 3 in Bell and Culshaw, 2001). Primary and secondary compression are similar to that of saturated clay. Primary settlements generally occur rapidly, with much of the settlement occurring during the actual application of load. Loess also may exhibit creep deformation under loading.

Northmore *et al.,* (1996) noted that brickearth from Essex exhibited rapid consolidation with coefficients of vertical consolidation of between 72 to 94 m^2/yr being found (cf. Table 3 in Bell and Culshaw, 2001). In many instances all the primary consolidation may take place within a half minute of loading, which is essentially showing collapse of the soil fabric.

Loess has a much higher vertical than horizontal permeability, which is enhanced by long vertical 'voids' in the loess structure that are formed by fossil rootholes and vertical fissures. Because of this, deposits of loess are better drained (their permeability ranges from 10^{-5} to 10^{-8} m s^{-1}) than true silts.

Loess/brickearth is also a highly erodable material. This was observed with catastrophic consequences with the collapse of the Teton Dam in USA (Smalley and Dijkstra, 1991)

Collapse potential

Soils such as loess, brickearth and certain other wind-blown silts may have the potential to collapse when wetted or wetted under loading. This process is now frequently referred to as 'hydro-consolidation' (Rogers *et al.,* 1994). The structural stability of loess soils has been related not only to the origin of the material, to its mode of transportation and to depositional environment, but also to weathering. Furthermore, in more finely textured loess deposits, high capillarity potential plus high perched groundwater conditions have caused loess to collapse naturally through time, thereby reducing its porosity. This reduction in porosity, combined with a high liquid limit, makes the possibility of further collapse less likely. Gao (1988) concluded that highly collapsible loess usually occurs in regions (primarily major river valleys) near the source of the loess where its thickness is at a maximum and where the landscape and/or the climatic conditions are not conducive to development of long-term saturated conditions within the soil. A good review of collapse in loess soils has been provided by Rogers *et al.,* (1994).

Collapse of the microstructure of loess occurs when the soil is flooded under loads exceeding its natural overburden pressure. Many criteria have been proposed in the assessment of collapse potential, these being based on soil properties such as natural moisture content, void ratio or index properties (see Rogers *et al.,* 1994 and Northmore *et al.,* 1996). These can be misleading as they are often based on remoulded and approximate soil properties, and inappropriate evaluation can occur (Northmore *et al.,* 1996).

Generally, an index collapse test is a better indication of collapse potential, especially as the total stresses at the point of collapse can be measured. Jennings and Knight (1975) developed the double oedometer test for assessing the response of a soil to wetting and loading at different stress levels (that is, two oedometer tests are carried out on identical samples, one being tested at its natural moisture content, whilst the other is tested under saturated conditions with the same loading sequence being used in both cases). The essence of this is illustrated in Figure 5. Jennings and Knight subsequently modified the test so that it involved loading an undisturbed specimen at natural moisture content in the oedometer up to a given load. At this point the specimen is flooded and the resulting collapse strain, if any, is recorded. The specimen is then subjected to further loading. The total consolidation upon flooding can be described in terms of the coefficient of collapsibility (C_{col}). This basic technique is still in use (see Figure 5). However, at best it should only be considered as an index test due to difficulties of sampling and control of effective stresses. Various modifications have been suggested, e.g. Houston *et al.,* (1988). Although only approximate, these tests do give a repeatable and reproducible, qualitative indication of collapse. Often it is the collapsibility risk that is more important to assess than the actual amount of collapse that will occur. Once collapse risk/severity has been assessed then a suitable ground treatment can be designed. Bell and Culshaw (2001) provide further discussion of both collapse severity assessment and possible treatment methods available for collapsible loess soils.

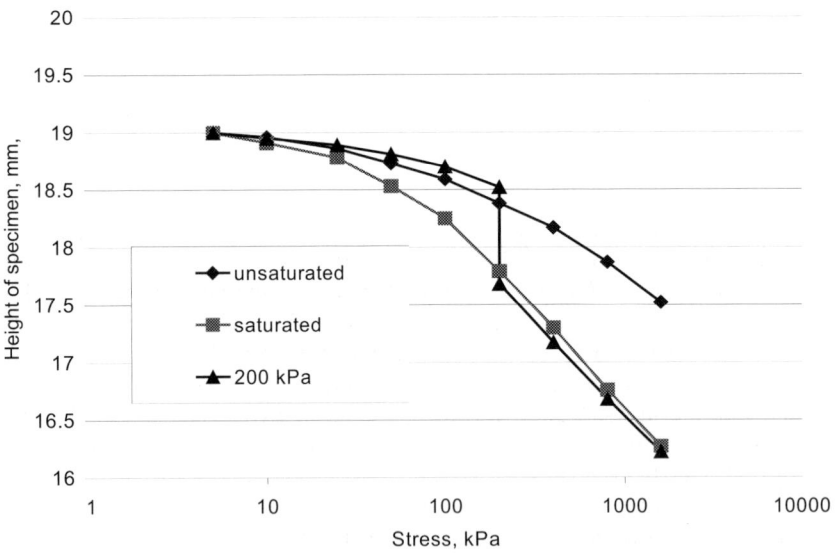

Figure 5 An example set of index oedometer loading curves on brickearth from Star Lane Brickworks Essex, showing collapse at 200kPa on inundation (after Miller *et al.,* 2001)

The oedometer test was also used to assess the degree of collapsibility of brickearth samples from south Essex, the results from such tests are illustrated in Figure 6 (after Northmore *et al.*, 1996).

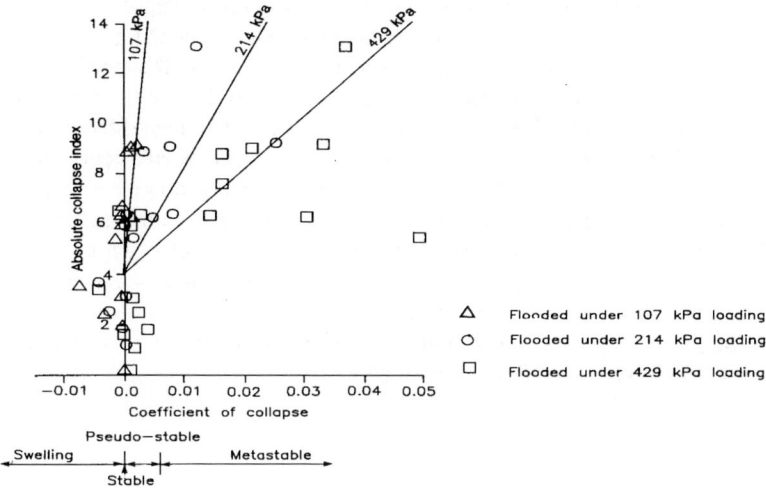

Figure 6 An illustration of collapse criteria applied to oedometer data for brickearth from south Essex (after Northmore *et al.,* 1996)

In Figure 6 the total stress at collapse is indicated and hence, the degree of collapsibility has to be qualified in terms of the level of loading at which flooding takes place. For instance, soils that were metastable when flooded at 429kPa were not necessarily metastable when flooded at lower loads. A distinction is drawn between metastable and pseudostable soils. Metastable soils are regarded as those which undergo more than 1% collapse on flooding when subjected to the highest loading (that is, 429kPa) whereas pseudostable soils exhibit less than 1% collapse under the same load. The degree of the collapsibility of the brickearth from south Essex varied both laterally and with depth (Northmore *et al.,*1996). It should be noted that one sample had a negative value indicating that it swelled when wetted. This is not untypical at lower mean normal effective stresses and where swelling clay minerals such as montmorillonite are present in the clay matrix. Derbyshire and Mellors (1988) have made similar observations.

Northmore *et al.,* (1996) observed that, upon flooding in the oedometer, certain specimens of brickearth became saturated almost instantaneously, that is, there was a rapid intake of water into the pore space. The rapidity with which this saturation occurred may be due to the rapid breakdown of bonds between silt particles giving rise to a possible 'suction effect'. The rapid inflow of water in the direction of flooding may lead to a re-orientation of the soil fabric which enhances the free flow of water along the drainage paths, that is, vertically within the specimen, consequently causing an increase in permeability. Such a mechanism may explain the rapidity with which collapse occurs when some specimens of brickearth are flooded under load.

Figure 6 shows the relation between the 'absolute collapse index' and the coefficient of collapse. It would appear that values of 'absolute collapse index' greater than 6 would predict metastable collapse of the brickearth in south Essex beneath a moderately heavy construction load (that is, greater than 200kPa), while values greater than 5 would indicate potential metastable collapse beneath foundations subject to heavy structures (that is, greater than 400kPa). The 'absolute collapse index' is defined as:

$$i_{ac} = m/(Sr - PL) \hspace{2cm} (1)$$

where,   m is the natural moisture content
Sr is the degree of saturation
PL is the plastic limit

Although loess collapse is a less significant problem in the UK relative to the rest of the world, in recent years it has caused damage with financial repercussions (Cattell 1997). Structures exerting loads in excess of 200kPa are likely to result in collapse settlements if constructed on many UK loess soils. However, even with lower loads from low-rise buildings, UK loess deposits have the potential to collapse if saturated, for example, by a leaking drain.

It should be noted that collapse potential is a function of moisture content and degree of saturation. At higher moisture contents (and associated higher degrees of saturation) partial collapse of loess may occur. Saturations of around 70 to 80% have been shown to remove a significant degree of collapsibility (Miller *et al.,* 2001). This could explain in part why the collapse severity in the UK brickearth is relatively low by world standards. This is associated both with the reworking that many brickearth deposits have experienced in their history and the wetter maritime climate enjoyed by the UK. Bell *et al.,* 2001 and others present results that illustrate the dramatic change from stronger brittle shearing behaviour to a weaker plastic type of behaviour that brickearths exhibit as moisture contents increase. This change can occur with only a relative small increase in moisture content, typical of soil of low plasticity indexes.

Case history

There are numerous cases around the world where loess collapse on inundation, when under load, has resulted in damage and considerable cost. It has been estimated that in countries like Bulgaria, loess collapse has cost some US$30 million over recent years (Karastanev, D. *pers. comm.*). As such, loess collapse can be considered a major global geohazard to the built environment. In the UK cases are much rarer. However, with increased development, and the presence of isolated pockets of loess material across southern Britain, loess collapse has the potential to surprise.

An example of this has been discussed by Cattell (1997), who noted that approximately one third of sites he investigated in Devon where distress to low-rise building had been observed, had loessic silt layers present. Subsidence problems are common in south Devon in areas underlain by Permian Breccias, and are often worst in areas where loess is present (Cattell 1997). In one area where a continuous sheet of loessic soil is present in a shallow dry valley, all the houses underlain by this deposit have foundation related problems to varying degrees, with three properties requiring either substantial underpinning or demolition (Hunter *et al.,* 2001). Figure 7 shows the level of damage that can occur, with the property illustrated having been subsequently demolished. This property had suffered both general subsidence of load bearing walls, and localized subsidence from leaking drains. The sub-soils are colluvial sandy silts with significant amounts of loess (Hunter *et al.,* 2001). The movement of the foundation occurred in two stages with longer term settlement of loose loessic colluvium, followed by a rapid localized effect associated with the leaking drains. This would have resulted in collapse if the sub-soil had still been metastable. However, it is likely that the additional moisture significantly softened the loessic sub-soils and hence significantly enhanced the settlements observed (further details are given in Hunter *et al.,* 2001).

Figure 7 Subsidence induced cracking in a house in Devon, as a result of loess collapse and subsequent post-wetting consolidation (after Hunter *et al.,* 2001)

Other examples where brickearth may be encountered by engineers include the development of building and infrastructure facilities in and around London. It was, for example, the designation of a development area eastwards of London into south Essex to accompany the construction of a proposed third London airport that initiated the investigations presented by Northmore *et al.* 1996, on the brickearth of Essex. Moreover, as landfill space in the southeast of England becomes scarcer, ex-brick pits where brickearth has been extracted could be used for waste disposal. This would present some interesting engineering challenges to ensure that full geotechnical designs to every increasing landfill standards are met.

Conclusions

Loess is a major worldwide geohazard to the built environment. Although it is a much less serious problematic soil in the UK, challenges are and potentially will be faced by engineers in Britain. Loess in the UK (otherwise most commonly known as brickearth) occurs in significant thickness across Essex, Kent and Sussex, with many more localised deposits across much of southern Britain. Loess soils in the UK are typically pale yellow-brown soils consisting of silt-sized quartz with clay minerals and carbonates. They are of aeolian origin, although they may have experienced reworking during their geological history, since their original deposition. The secondary minerals within the loess fabric typically bridge the quartz particles enabling a metastable open structure to form, which upon inundation (normally when under load) causes collapse to occurs, a process known as hydro-consolidation. This collapse associated with the relatively high vertical permeability of loess derived from the long void structures formed by fossil root holes and vertical fissures, can be rapid. There are a number of ways to examine the collapse and to assess the severity of the collapse. The most common, and currently the most reliable procedure is to undertake double oedometer or modified single oedometer collapse testing using carefully acquired undisturbed tests samples. Although, this will not accurately give the absolute amount of collapse in all field situations, these index tests are extremely effective at indicating the potential risk of collapse and as such can provide a useful tool when engineering loess soils. Collapse criteria based on soil index properties such as natural moisture content, void ratio and consistency limits, although useful, may only be locally applicable and provide a rough indication of whether or not a particular loess soil may be collapsible. In any case, such criteria can only be derived by correlation with oedometer collapse tests in the first instance.

References

Bell, F.G. and Culshaw, M.G. 2001. *Problematic soils: A review from a British perspective.* In: Problematic Soils, (Jefferson, I., Murray, E.J., Faragher, E. and Fleming, P.R. eds.) Thomas Telford Publishing, London. (*in press*)

Bell, F.G., Culshaw, M.G. and Northmore K.G. 2001. *A review of the Geotechnical properties of some collapsible soils with a particular emphasis on brickearth.* In: Silt and Siltation: problems and solutions., (Jefferson, I., Rosenbaum, M.S. and Smalley I.J. eds.) Thomas Telford Publishing, London, *(in press)*

Boardman, D.I., Rogers, C.D.F., Jefferson, I. & Rouaiguia, A. 2001. *Physico-chemical characteristics of British loess.* In: Proceedings of the 15th International Conference on Soil Mechanics and Geotechnical Engineering, Istanbul, Turkey, *(in press).*

Catt, J. A. 1977. *Loess and coversands.* In: British Quaternary Studies, Recent Advances, (Shotton, F. W. ed.). Clarendon Press, Oxford, 221-229.

Catt, J. A. 1985. *Particle size distribution and mineralogy as indicators of pedogenic and geomorphic history: examples from soils of England and Wales.* In: Geomorphology and Soils, (Richards, K. S., Arnett, R. R. and Ellis, S. eds.) George Allen and Unwin, London, 202-218.

Catt, J.A. 1988. *Quaternary geology for scientist and engineers.* Ellis Horwood Limited, Chichester.

Catt, J.A. & Staines, S.J. 1982. *Loess in Cornwall.* Proc. Usshers Soc., 5, 368-375.

Cattell, A.C. 1997. *The development of loess-bearing soil profiles on Permian Breccias in Torbay.* Proc. Usshers Soc., 9, 168-172.

Derbyshire, E., Dijkstra, T.A., & Smalley, I.J. (eds) 1995. *Genesis and properties of collapsible soils.* Series C: Mathematical and Physical Sciences – Vol. 468. Kluwer Academic Publishers, Dordrecht.

Derbyshire, E. and Mellors, T. W. 1988. *Geological and geotechnical characteristics of some loess and loessic soils from China and Britain.* Engineering Geology, 25, 135-75.

Dibben, S.C., Jefferson, I.F. and Smalley, I.J. 2001. *A microstructural computation simulation model of loess soils.* In Civil-Comp, Vienna, *(in press)*

Eden, D.N. 1980. *The loess of north-east Essex, England.* Boreas, 9, 165-177.

Fookes, P. G. and Best, R. 1969. *Consolidation characteristics of some late Pleistocene periglacial metastable soils of east Kent.* Quarterly Journal of Engineering Geology, 2, 103-128.

Foster, S.S.D., Morigi, A.N. & Browne, M.A.E. 1999. *Quaternary geology – towards meeting user requirements.* British Geological Survey, Nottingham,

Gao, G. 1988. *Formation and development of the structure of collapsing loess in China.* Engineering Geology, 25, 235-45.

Gibbons, W. 1981. *The Weald.* Unwin Paperbacks, London.

Houston, S. L., Houston, W. L. and Spadola, D. J. 1988. *Prediction of field collapse of soils due to wetting.* Proceedings of the American Society of Civil Engineers, Journal of the Geotechnical Engineering Division, 114, 40-58.

Hunter, J.M., Jefferson, I.F., Cattell, A.C. & Smalley, I.J. 2001. *Loess in Devon – its origin, distribution and geotechnical properties.* Submitted to the Bulletin of Engineering Geology and the Environment.

Jennings, J. E. and Knight, K. 1975. *A guide to construction on or with materials exhibiting additional settlement due to collapse of grain structure.* In: Proceedings of the 6th African Conference on Soil Mechanics and Foundation Engineering, Durban, 99-105.

Karastanev, D. 2001. *Pers. comm.*

Kerney, M. P. 1965. *Weichselian deposits in the Isle of Thanet, east Kent.* Proceedings of the Geologists' Association, 76, 269-274.

Lee, M.P. 1979. *Loess from the Pleistocene of the Wirral Peninsula, Merseyside.* Proceedings of the Geologists' Association, 90, 21-26.

Lill, G.O. & Smalley, I.J. 1978. *Distribution of loess in Britain.* Proceedings of the Geologists' Association, 89, 57-65.

Miller, H., Jefferson, I.F., Djerbib, Y. and Smalley, I.J. 2001. *The collapse of Quaternary metastable loess.* Quarterly Journal of Engineering Geology and Hydrogeology. (in press)

Northmore, K. J., Bell, F. G. and Culshaw, M. G. 1996. *The engineering properties and behaviour of the brickearth of south Essex.* Quarterly Journal of Engineering Geology, 29, 147-161.

Northmore, K.J. & Hobbs, P. 1999. *Engineering properties of quaternary deposits.* Introductory course on Quaternary processes, course notes, British Gelogical Survey, Nottingham.

Pye, K. 1987. *The nature, origin and accumulation of loess.* Quaternary Science Reviews, 24, 653-667.

Roberts, M.C. 1985. *The geomorphology and stratigraphy of the lizard loess in south Cornwall, England.* Boreas, 14, 75-82.

Rogers, C. D. F., Djikstra, T. A. and Smalley, I. J. 1994. *Hydroconsolidation and subsidence of loess: studies from China, Russia, North America and Europe.* Engineering Geology, 37, 83-113.

Rogers, C.D.F. & Smalley, I.J. 1993. *The shape of loess particles.* Naturwissenschaften, 80 (1), 461-462.

Rose, J., Lee, J.A., Kemp, R.A. & Harding, P.A. 2000. *Palaeoclimate, sedimentation and soil development during the last glacial stage (Devensian), Heathrow Airport, London, UK.* Quaternary Science Reviews, 19, 827-847.

Smalley, I.J. & Dijkstra, T.A. 1991. *The Teton Dam (Idaho, USA) failure: problems with the use of loess material in earth dam structures.* Engineering Geology, 31, 197-203.

Shrinking and swelling of clays

R Driscoll and R Chown
BRE, Ganston, Watford, Herts, WD25 9XX

Introduction
The shrinking of clay soils as a result of evaporation and transpiration by vegetation is a common cause of damage to low-rise buildings in the UK; the insurance industry regularly attracts annual domestic building damage claims in the order of £400 million (Driscoll and Crilly, 2000), (Figure 1).

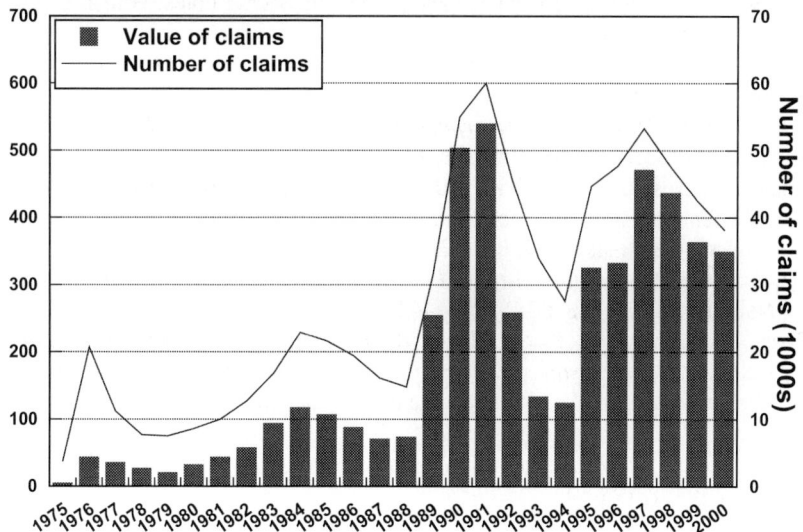

Figure 1 Growth of subsidence insurance claims

This is a particular problem in the south-east of England, where there are widespread outcrops of high plasticity clays (such as the London and Gault Clay) and where the summer is warmer and drier than in the rest of the country. These factors, coupled with the influence of trees, can result in relatively deep drying, or *desiccation,* of the soil to a considerable extent around trees:

Problematic Soils. Thomas Telford, London, 2001

desiccation extending to 5 m below ground level to distances of up to 15 m from trees is not uncommon. Rehydration of desiccated clay, seasonal near the surface, and long-term at depth following tree removal and the cessation of transpiration, may lead to prolonged swelling and building uplift. Cheney (1988) recorded about 160mm of heave of a single-storey building over a 25 year period.

Shrinking and swelling processes are controlled by changes in the effective stresses in the soil. For low-rise buildings that impart only small total stress changes to the ground, changes in pore water suction dominate the volume changes processes. Suctions exceeding hydrostatic values in soil above a water table are deemed 'desiccation'; such values are induced primarily by evapo-transpiration involving moisture extraction through the roots of vegetation; occasionally, desiccation may be induced by elevated temperatures such as occur beneath furnaces.

BRE (1996) has published Digest412 which identifies and assesses four main techniques for detecting desiccation:

- comparisons of soil water contents with soil index properties;
- comparisons of soil water content profiles;
- comparisons of strength profiles; and
- effective stress or suction profiles.

Digest 412 concludes that the first of these techniques may often give misleading results and that the others, though generally more expensive, are preferable. Because each technique has its own disadvantages, it is best not to rely on one method. It is also concluded that an assessment of the effective stress or suction profile within the ground may give the most fundamental indicator of the state of desiccation within the profile.

The filter paper method of suction determination (Chandler *et al.* 1992; Crilly and Chandler (1993)) provides the most readily available and practicable means of determining a suction profile, though recent developments in equipment for the measurement of soil suction (Ridely and Burland (1993)), may provide more accurate, quicker and cheaper measurements.

Trees as a source of desiccation

BRE has for many years been studying the interaction between trees, clay soils and various types of foundation at a high plasticity London Clay site near Chattenden, Kent (Freeman *et al.*, 1991 and Crilly *et al.*, 1992). Part of the site was covered with 20 m - 25 m tall Lombardy Poplar trees which, following detailed observations of seasonal variations in ground and foundation levels and water contents, were felled in September 1990 to investigate the effects of ground swelling on some dummy foundations; these included pad foundations with a variety of loadings, a trench fill foundation split into sections with different types of protective, compressible material on each section, and 12 m long piles, instrumented with vibrating-wire strain cells to measure the distribution of loads in the piles.

Figure 2 shows the vertical ground movements (before and after tree removal) measured at various depths in the ground in an area under grass cover, remote from the influence of trees (Group 1), and in an area surrounded by trees (Group 2).

It is clear from these data that, many years after tree removal, ground swelling is far from complete; this is particularly evident from the ground movements at 2 m below ground level.

Examination of these and other results from the site leads to the following conclusions about movement:

1. Seasonal movements under grass cover are generally restricted to the top metre of soil. However, during 1990 a movement of 13 mm was recorded at 1 m below ground level. Given the high plasticity of the clay at the site and the extreme weather conditions at the time, it is considered that such a figure may form a reasonable upper bound to the level of movement that may occur under grass cover in the UK. In practice, *differential* movements would be less than this, and certainly within the range that a well-built low-rise property should withstand without cracking. There is therefore no evidence that the currently recommended minimum foundation depth of 0.9 m - 1.0 m is inadequate for construction in open ground.

2. Near trees, movements are clearly considerably larger and extend to much greater depths. In the area where measurements have been made, NHBC Standards (2000) would recommend a foundation depth of 3m. At this depth seasonal movements of up to 18mm magnitude were measured; this suggests that NHBC guidelines are adequate to prevent damage from seasonal movements. There has, as yet been little significant swelling at this depth at the location of the felled trees.

3. Following tree removal, recovery of the soil between ground level and 1 m below ground level took around six months. In contrast, it took around four years for the soil between 1 m and 2 m to recover.

4. Comparison of measured strains in the ground with measured water content changes shows that the 'water shrinkage factor' (as described by BRE (1996)) varies markedly with depth, both seasonally and following tree removal, knowledge of which leads to better estimates of vertical ground movement.

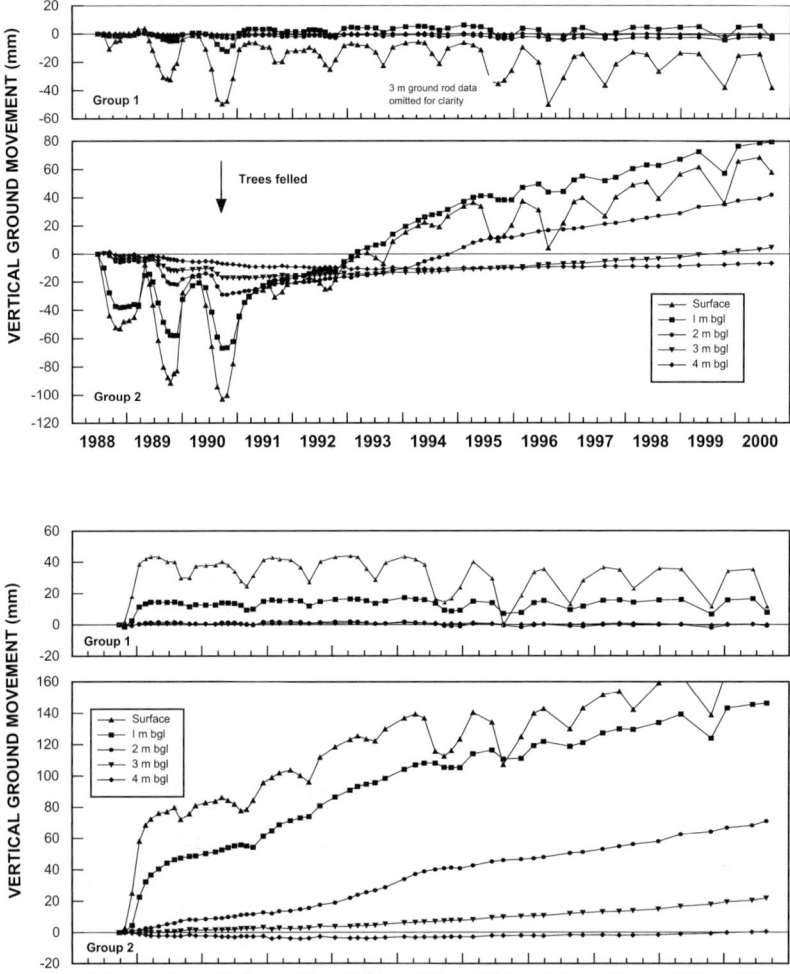

Figure 2 Ground movements at Chattenden. The upper plot shows results obtained since the first movements in June 1988. The lower plot shows an enlarged scale with results obtained since the trees were felled

5. In terms of foundation movements and soil drying, the effect of two consecutive dry summers (such as 1989 and 1990), with a dry

intervening winter, is considerably more severe than that of either summer individually, and clearly worse than the dry summer of 1995. This fact has implications for any potential impact from global climate change.

Dealing with trees on clay sites

Foundations for new buildings

The traditional forms of house building in Britain tend to produce a brittle response to any foundation movement, leading too readily to cracking, albeit usually of only minor aesthetic significance. Consequently, domestic building on shrinkable clay soils focuses on the construction of a foundation at a depth where only small ground movements are likely to occur. Traditional techniques are used that usually require little engineering input.

'Trench-fill' foundations

In the absence of proximate trees, there is ample evidence that foundations in clay soils need not exceed 0.9 – 1.0 m. depth, even in drought conditions when any slight subsidence movement should not affect the building's structure or utility. Careful observations of ground movement at Chattenden (Crilly *et al.*, 2000), where the clay is of very high potential soil volume change, have confirmed this.

To cater for the presence of trees, the NHBC Standard (2000) contains detailed charts of required foundation depths, depending on soil volume change potential and tree 'water demand', (mature) height and proximity (Figure 3).

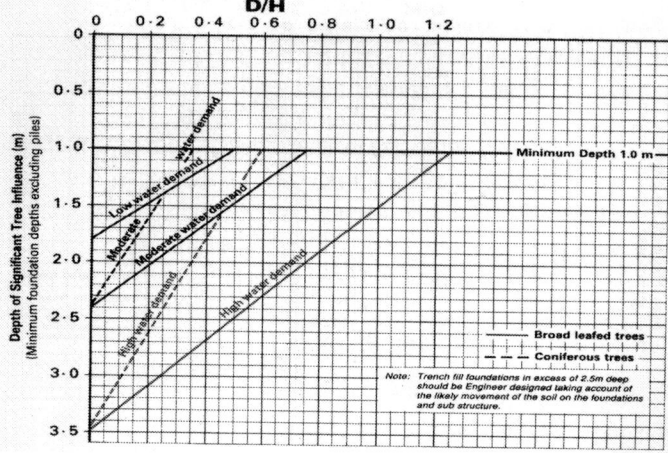

Figure 3 NHBC Standard 4.2 chart of foundation depth for 'high shrinkability' soils.

It is apparent that the NHBC charts have tended to lead to progressively deeper trench-fill foundations. BRE (1999) has highlighted the risk posed to these foundations by ground movements and recommends the use of small diameter pile foundations in proximity to trees. In fairness to the NHBC, their Standard (2000) suggests that foundations deeper than 2.5m should be 'engineer-designed'; however, this is accompanied by recommendations for foundation depths of up to 3.5m, implicitly suggesting deep trench fill.

Deep trench fill foundations require little design input and are readily excavated by back-hoe and filled with concrete. However, BRE (1999) have for many years argued that there are significant disadvantages in their use;

- The deep trench fill is likely to be more expensive than a bored pile solution, except for small schemes.
- Instability of the trench sides can lead to serious construction difficulties.
- There is a danger of horizontal and rotational foundation movement when large vertical areas of concrete are subjected to differential lateral pressures and shear forces resulting from either shrinking or swelling of the clay.

In addition, the construction of deep trench fill foundations produces large quantities of spoil. The foundation itself is vulnerable to poor workmanship and to poor construction detailing:

- concrete over-spill, or over-break in the excavations can result in unanticipated vertical forces being transmitted to the foundation
- some lateral pressure will be transmitted through compressible materials if the soil swells – therefore, vertical or near-vertical cold joints in the foundation concrete should be avoided
- missing or incorrectly placed protective compressible material will make the foundation susceptible to lateral and/or vertical movements.

Trench fill remains the foundation of choice except on very high potential clay, where trees have been removed, when builders are tending to adopt pile foundations to counter the threat of desiccation rehydration, swelling and foundation heave. There are several reasons for this reluctance to use piles for other than the most hazardous circumstances, not least being the absence of simple guidance for the design of such a foundation that does not lead to an overly long and therefore more expensive pile.

Pile foundations
As mentioned above, 12m long, reinforced, cast-in-place concrete piles, instrumented with strain cells to determine the distribution of load down the piles, have been monitored at Chattenden to record the impact of ground heave following tree felling and desiccation rehydration. Figure 4 shows the

desiccation in terms of reduced soil moisture content and increased soil moisture suction. profiles recorded close to and remote from trees.

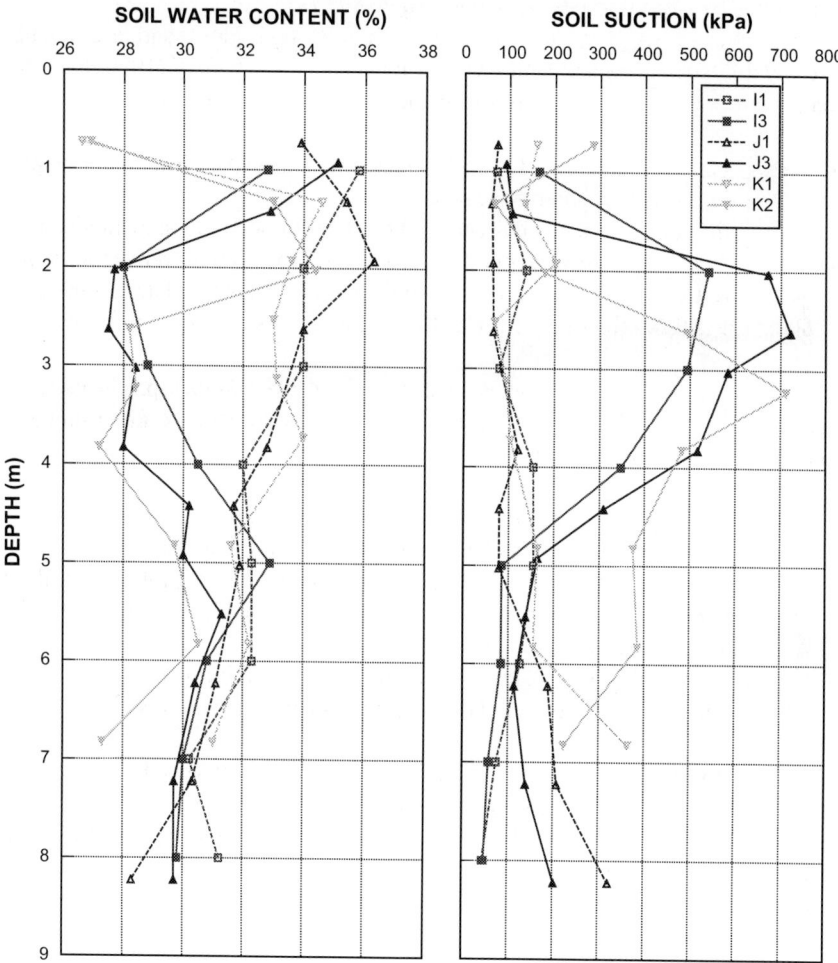

Figure 4 Evidence of desiccation at Chattenden

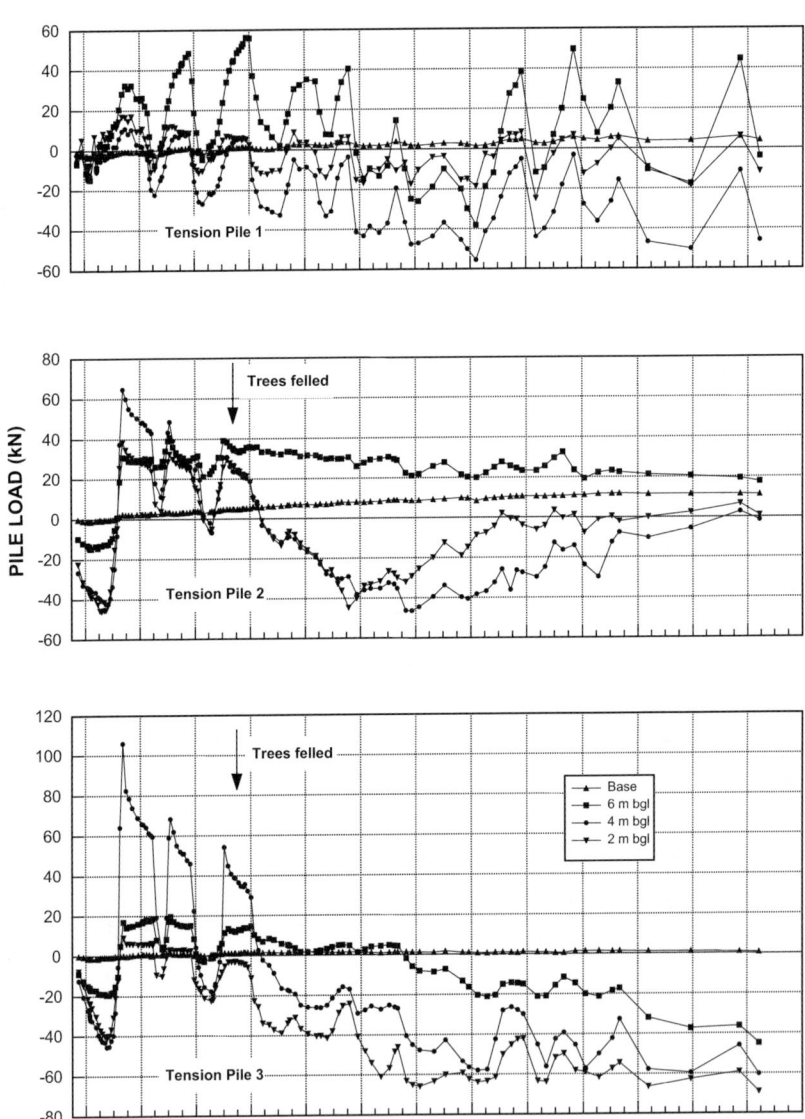

Figure 5 Axial forces in piles at Chattenden

Figure 5 shows axial forces measured in two instrumented piles installed within the area where trees were subsequently felled to induce swelling, while Figure 6 shows the corresponding shear stresses calculated to be acting piles 2 and 3 (pile 1 being 'open' ground).

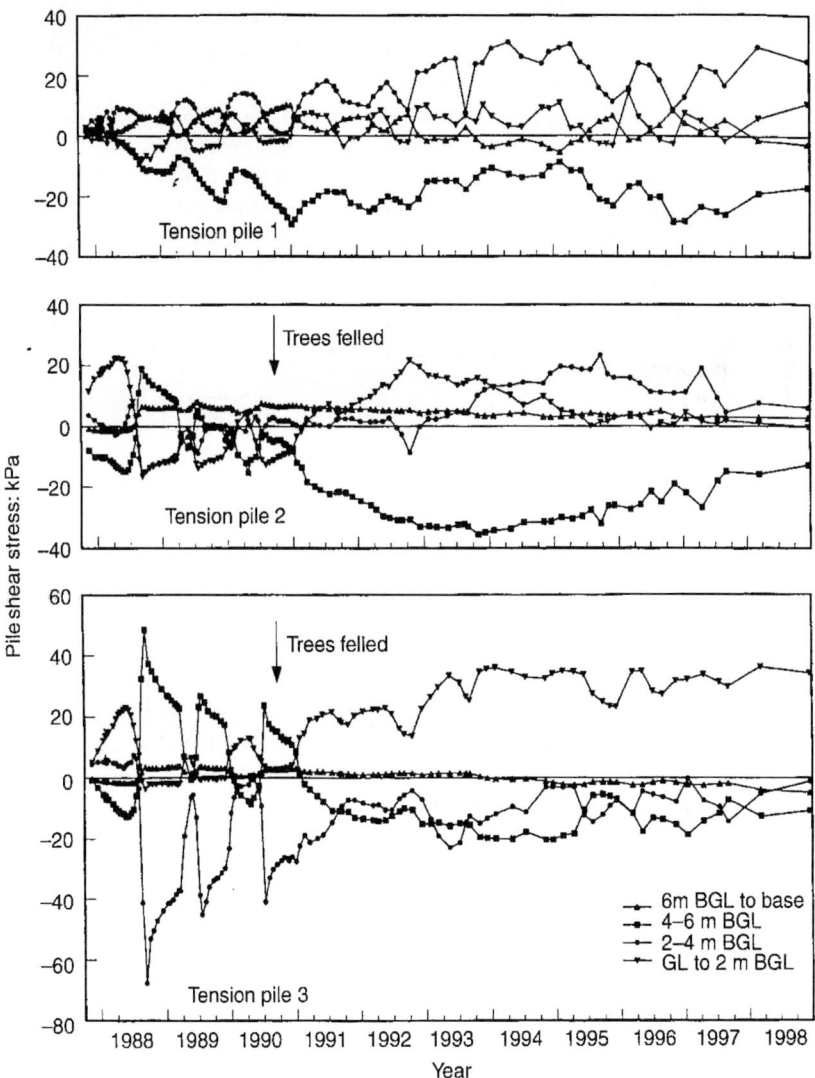

Figure 6 Stresses on pile shafts at Chattenden.

These and other measurements were used to assess methods for the design of piles to resist ground heave. It is suggested (Crilly and Driscoll, 2000) that:

- In a total stress design, $\alpha = 0.45$ is used on undesiccated strength values throughout the anchorage zone and that the greater of 0.6 times the undesiccated strength or 0.45 times the desiccated strength be used throughout the swelling zone. In this, it is assumed that strengths are measured in quick, undrained triaxial tests on 38mm. dia. samples; different values may be more appropriate for other test types or sample sizes.
- In an effective stress design, where the main unknown is the value of σ'_{rf}, the radial effective stress on the pile shaft at the point of failure, it is recommended that $\sigma'_{rf} = \sigma'_{ho}$ in the anchorage zone (σ'_{ho}, the free-field horizontal effective stress, may be estimated from K_0 and σ'_{vo} profiles). In the swelling zone, σ'_{rf} may be taken as the larger of either σ'_{ho} - calculated assuming that an increase in suction ($\Delta p'$) gives an increase in σ'_h above the at-rest value from

$$\Delta \sigma'_h = \frac{3v'}{1+v'} \Delta p'$$

or the maximum (undesiccated) horizontal stress obtained from, ay, published K_0 profiles.

Additional design points specific to piles in swelling ground that may need to be considered are:

- the application of dead load before significant uplift forces develop;
- owing to shrinkage of ground away from the piles, exclusion of the top 1-2m from the calculation of resistance to compressive load, especially if trees are retained nearby. Of course, the top 1-2m should be included in the uplift calculation.

Alternative foundation solutions

In other parts of the world, particularly the USA, South Africa and Australia, the use of stiff rafts on expansive clay soils is widespread. Their use in the UK is largely confined to deep, filled and soft, compressible sites where stable foundations would be prohibitively expensive.

Rafts
Stiff rafts are an NHBC-approved house foundation solution for shrinkable clay sites but they are not commonly used. The raft is intended to eliminate building deformation but does not prevent tilting. Unlike cracking, for which 'acceptability' criteria have been defined, there is no consensus on levels of 'acceptable' tilting; this has led to demolition of tilted but un-cracked buildings

in some instances. Guidance for the provision of stiff rafts in the UK is overly complex and onerous, and requires significant engineering input with a need for on-site skills in site preparation and reinforcement construction. These reasons are considered to be major stumbling blocks to their use. The publication of standard foundation types similar to those in the USA (Federal Housing Administration, 1968) for example, may stimulate their use.

Modular foundations
These novel foundations comprise elements that are prefabricated off-site and assembled on-site (Figure 7), often with post-tensioning to improve stiffness.

Figure 7 Modular foundation being installed at experimental site

The benefit of these alternative foundations is principally in their improved performance (due to increased stiffness) over traditional trench-fill and shallow strip foundations. The speed with which modular foundations can be constructed is significantly better than traditional alternatives. There are also significant improvements in the form of reduced waste, manufacturing scale tolerances and the potential in some cases to use the foundations for temporary structures, removing them once their purpose has been fulfilled.

Remedial activity in cases of tree-induced damage.

Tree removal or reduction
Surprisingly little is known about the effects on damaged buildings of tree removal or reduction; this stems largely from the tendency to underpin foundations as well as dealing with the tree. BRE monitored the impact of tree reduction on terraces of houses in the London Borough of Westminster. In Figure 8, precise level monitoring data are shown for points on the façade of a terrace of 10 two-storey dwellings in a street with 25 trees located at regular spacings in the pavements. The impact of some reduction on movements is clearly visible as is the fact that biennial pruning has not been particularly successful in reversing the impacts of early subsidence as the trees grew.

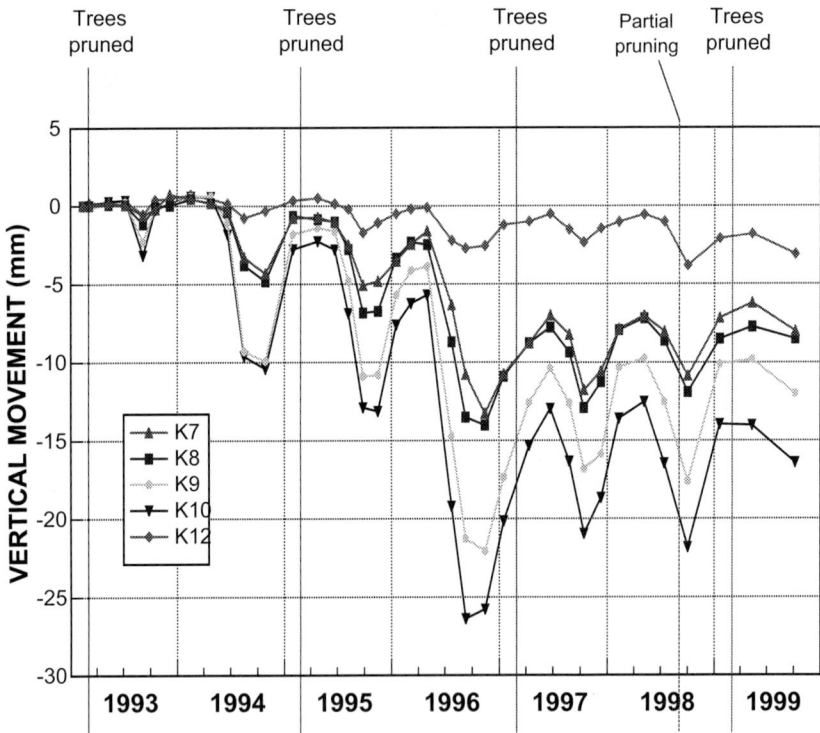

Figure 8 Vertical movement of houses in London Borough of Westminster

Underpinning
The various forms of underpinning to deepen unstable foundations on shrinking and swelling clays (Hunt *et al.*, 1991) have evolved as a direct result of the foundation types used in the UK. Whilst a small number of modern properties,

built in the last decade or two, suffer subsidence damage, the bulk of the domestic subsidence claims arise in housing stock that is approximately 50 years old or older and founded on shallow, unreinforced strip foundations. This is confirmed in the BRE subsidence database (Crilly and Chandler, 1993) where the most common foundations implicated in cases of damage recorded in the database were strip or brick footings. Information from the database indicates that mass concrete was the most common form of underpinning and was applied predominantly to only a part of the building. There is growing evidence that properties that have been previously underpinned are more likely to suffer a recurrence of damage than properties that have not been previously underpinned.

Conclusions
Swelling and shrinking of clay soils, and their potentially damaging impact on the foundations of low-rise buildings are caused by changes in suction in the pore water of the soil. These changes are produced largely by the evapotranspiration of water by trees. Detection of this desiccation is reasonably well dentifiable but estimation of the amounts of movement are not very reliable.

Foundation provision for new buildings has tended to be rather conservative and to use traditional techniques that do not always ensure a damage-free building, in the long turn. Improvements to the design of piles should make them a more cost-effective and technically superior alternative. The provision of stiff rafts, popular in some other countries, may be hindered by overly complex and onerous design requirements and concerns about overall building tilt, rather than deformation leading to cracking. The use of innovative foundations may bridge this gap, offering 'off-the-shelf ' approved systems that provide technical improvements and reductions in time and waste.

Remediation in cases of damage tends too often to involve partial underpinning in conjunction with action to remove or reduce tree influence. There is growing evidence that partially underpinned buildings are prone to further differential movements many years later. Reduction in tree volume should be regarded as a temporary expedient. Significant improvements to underpinning practice (cost, time and disruption to the homeowner) are restricted by the foundation types traditionally used in the UK.

Acknowledgements
Much of the work described in this paper has been funded over several years by the Construction Directorate of the Department of the Environment, Transport and the Regions.

References

BRE. 1996. *Desiccation in clay soils.* Digest 412. CRC Ltd, Watford.

BRE. 1999. *Low-rise building foundations: the influence of trees in clay soils.* Digest 298. CRC Ltd, Watford.

Chandler, R J, Crilly, M S & Montgomery-Smith, G. 1992. *A low-cost method of assessing clay desiccation for low-rise buildings.* Proc Inst. Civ Eng Civ Engg, 92, 82-89.

Cheney, J E. 1988. *25 years' heave of a building constructed on clay after tree removal.* Ground Engng, 21, (5), 13-27.

Crilly, M. S, Driscoll, R. M. C. & Chandler, R. J. 1992. *Seasonal ground and water movement observations from an expansive clay site in the UK.* Proc 7th Int. Conf Expansive Soils, Dallas, 1, 313-318.

Crilly, M. S. & Chandler, R. J. 1993. *A method of determining the state of desiccation in clay soils.* BRE Information Paper No 4/93.

Crilly. M. S. & Driscoll. R. M. C. 2000. *The behaviour of lightly loaded piles in swelling ground and implications for their design.* Proc. Instn Civ. Engng, 143, 3-16.

Driscoll, R. & Crilly, M. 2000. *Subsidence damage to domestic buildings: lessons learned and questions remaining.* FBE Report 1, CRC Ltd, Watford.

Federal Housing Administration. 1968. *Criteria for selection and design of residential slab-on-ground.* BRAB Report No. 33, National Academy of Sciences, USA.

Freeman, T. J, Price, G. & Crilly, M. S. 1991. *Ground induced loading of a bored pile.* Proc 4th Int. Conf Ground Movements & Structures, Cardiff, 568-586.

Hunt, R., Dyer, R. H. & Driscoll, R. M. C. 1991. *Foundation movement and remedial underpinning in low-rise buildings.* BR 184. CRC Ltd, Watford.

National House Building Council. 2000. *Building Near Trees.* Standards Chapter 4.2. NHBC, Amersham.

Ridley, A. M. & Burland, J. B. 1993. *A new instrument for the measurement of soil moisture suction.* Geotechnique, 43, 321-324.

Behaviour of highly compressible clays and silts

E.R. Farrell
University of Dublin, Trinity College, Dublin, Ireland.

Introduction

Our basic understanding of the behaviour of soft ground under embankment loadings is relatively well developed and this has contributed to the design and construction methods that are used in practice. Relatively advanced methods of soil testing, combined with appropriate theoretical methods of analysis have been shown to give good predictions of soil strength and soil settlements. Complicated soil models can readily be incorporated into numerical methods to give greater sophistication in the design process.

The sophisticated methods of testing and analysis are appropriate for relatively uniform soils such as marine clays or lacustrine deposits. The genesis of certain deposits, however, result in soils which can be highly variable and consequently less amenable to the determination of appropriate geotechnical parameters from discrete point tests. For example, soft soils deposited on the margins of some of the large rivers in Ireland are a combination of silts, organic silts or silty peats which have properties that depend on the relatively recent geological and hydrological history of the location. Such soils are considered to represent a relatively random variation from alluvial to marshland conditions. In such situations it is necessary to appreciate the effects of these variations and to anticipate the global response of the deposit.

This paper presents the results of a case history of an instrumented site where fill has been placed relatively rapidly on deposits of alluvial/marshland soils. The ground conditions at the site along with the recorded settlements and water pressures are analysed. The *in situ* soil response is back figured from the recorded settlements and compared with that estimated from laboratory and field tests carried out prior to the placement of the fill. The findings at this site are compared with those back figured from similar sites in Ireland and recommendations regarding methods of determining the relevant soil parameters are presented.

Problematic Soils. Thomas Telford, London, 2001

The site and ground conditions

The site is that of the Waste Water Treatment Plant which is located on the flood plain of the River Shannon on the south of Limerick, Ireland. This project is part of the Limerick Main Drainage project which is to upgrade the drainage system for the city. The topography of the original site was relatively flat, which is to be expected, at about 1.5m OD (Malin). The site is on the south bank of the river and flood protection banks were constructed during the last century to prevent flooding. The tidal fluctuation of the river at this point is from about +2.7m OD to –2.8mOD with a mean of about 0m OD.

The ground conditions over the site comprise 5m to 8m of very soft organic silts with thin peat layers (<0.4m) and sometimes with shells. There was usually an upper 0.8m to 1m crustal layer. These soft alluvial/marshland soils overlay a relatively thin layer of gravely silt over glacial till and fluvial gravels.

The water pressure in the alluvial/marshland soils was between +0.5m to +1m OD which is about 0.5 to 1m below ground level.

General soil properties

The moisture contents of samples recovered from the ground investigation is shown plotted against depth on Figure 1. These were generally between 50 and 150% although some values of the order of 400% were recorded. This shows the variation in soil across the site and highlights the more organic layers. The organic contents determined by loss of ignition on five samples with moisture contents between about 68% and 140% were between 4.75% and 14%. The higher values were generally recorded on those samples with the higher moisture contents, which is to be expected. The liquid and plastic limits plot as Silts of high plasticity on the plasticity chart. The effects of the organics can mask the nature of the fine soil, however particle size distribution analyses on samples which had been treated with hydrogen peroxide to remove the organics recorded less than 20% with equivalent diameter below 0.002mm (the clay size fraction). The majority of the particles are thus in the silt size range. The natural moisture contents are close to the liquid limits as would be expected for a normally consolidated soil.

The relatively high moisture contents and the presence of organics necessarily implies low specific gravity of the soil particles and that the bulk density of the soil is low. This results in low initial effective stresses in the ground. The specific gravities recorded in the laboratory were between 1.31 and 2.46 which is consistent with the variation of organic content. There was a corresponding wide variation in the bulk density with an average value of about 1.43 Mg/m^3.

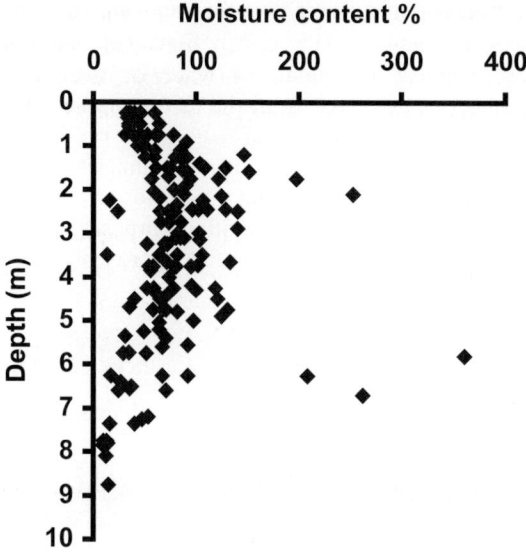

Figure 1 Moisture content versus depth

Undrained shear strength

The undrained shear strength was determined using the following approaches:-

a) Unconsolidated undrained triaxial tests (UU) on U100 samples
b) Unconsolidated undrained triaxial tests (UU) on piston samples
c) Isotropically consolidated undrained triaxial tests
d) Geonor H-70 in-situ vane
e) Geonor H10 penetration vane
f) CPTu
g) Direct simple shear

The undrained shear strength determined using the various approaches is necessarily different due to the differences in the stress path imposed on the soil. The Geonor penetration vane (H10) is considered to be a reliable method of determining the c_{uvane} and this method gave undrained strengths which were generally lower than those interpreted from the H 70 vane which is a cruder instrument. As expected, the undrained shear strengths from the U100 gave the lowest results (9kPa to 17kPa with an average of 13kPa when those from the crustal layer are excluded) whereas those from the piston samples were about 20kPa. The c_u values from undrained triaxial tests which had been initially consolidated to 10kPa were between 8 and 15kPa with an average of about 12kPa. The c_u from direct simple shear tests was about 13kPa which is, as expected from the different stress paths, lower than the vane and that from the undrained triaxial compression tests. The c_u/σ_{vc} from the isotropically consolidated triaxial compression tests was about 0.4 (where σ_{vc} is the effective vertical consolidation pressure) whereas that from the direct simple shear test was about 0.32.

A total of 12 No. CPTu probes were put down on the site and gave reasonably consistent results. Two of these were put down close to the location of the Geonor H10 penetration vanes test for calibration purposes. There was good agreement between the c_u determined from the two approaches for an N_k of 17 as can be seen from Figure 2 below, where

$$c_u = (q_t - \sigma_{vo})/ N_k \qquad\qquad (1)$$

where q_t is the cone resistance corrected for unequal area effects
              σ_{vo} is the total vertical stress
              N_k is the empirical cone factor (Lunne *et al.*, 1997)

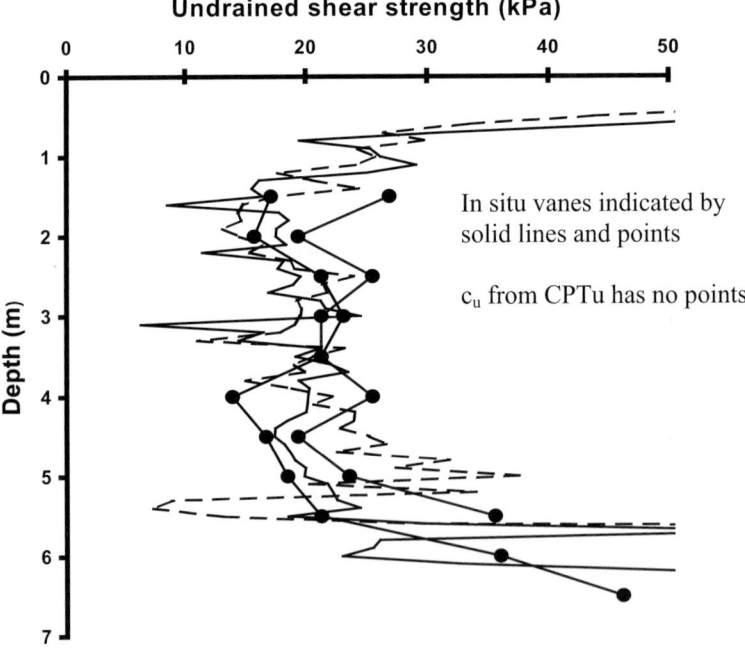

Figure 2 Comparison between field vane and c_u from CPTu with N_k=17

Effective stress parameters

The effective stress parameters determined in isotropically consolidated undrained triaxial compression tests were c' between 0 and 8.5kPa and ϕ' of 35.5 to 48°. These are relatively high values, but as discussed later, are typical of the parameters determined from triaxial compression tests on these soils. Lower values can be expected in direct shear tests.

Compression parameters

The settlement parameters were determined assuming a bi-linear e-log σ_v' plot with a slope of C_r below the preconsolidation/yield stress and a slope given by the compression index C_c above that point. Previous experience (Morris, 1998) has indicated that the reconstruction of the laboratory curves using Schmertmann's method (3) gives a reasonable estimate of the *in situ* behaviour under the primary consolidation phase. The preconsolidation/yield stress (σ_c') was interpreted using the Casagrande method and was estimated to be about 30kPa.

The principal parameters are the preconsolidation/yield pressure and the compression index C_c. The latter value varied considerably with moisture content and with depth and it is considered that a better parameter for use in this type of soil is the compression index ratio (CIR) which is defined as $C_c/(1+e_o)$. The use of the CIR gives a clearer pattern and is a better representation of the compression characteristics of these soils. The variation in this parameter with depth is indicated on Figure 3.

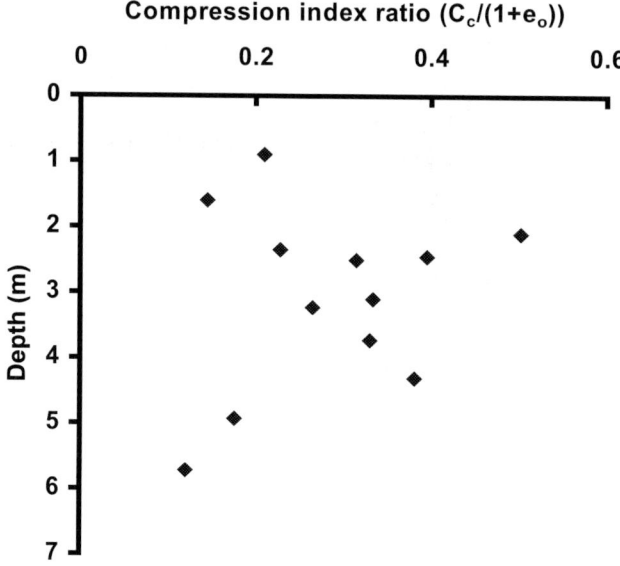

Figure 3 Compression index ratio versus depth

The parameters for the analysis of secondary consolidation (creep) were not investigated in detail on this project.

Coefficient of consolidation

The coefficient of consolidation (c_v) was determined from standard laboratory consolidation tests and in dissipation tests of the CPTu. Those values interpreted from the standard oedometer tests relate to a condition with vertical drainage (c_{vv}). Relatively high c_{vv} values below σ_c' (the preconsolidation/yield pressure) were interpreted, reducing at higher effective stresses as can be seen from Figure 4. The coefficient of consolidation interpreted from the CPTu tests, which would represent a greater proportion of horizontal dissipation but would represent the value below the yield pressure, was between 4 and 6 m²/yr. No constant rate of deformation tests (CRS tests) were carried out on this site although this method may offer advantages over the standard methods as it gives a continuous profile of the variation of the coefficient of consolidation with effective stress.

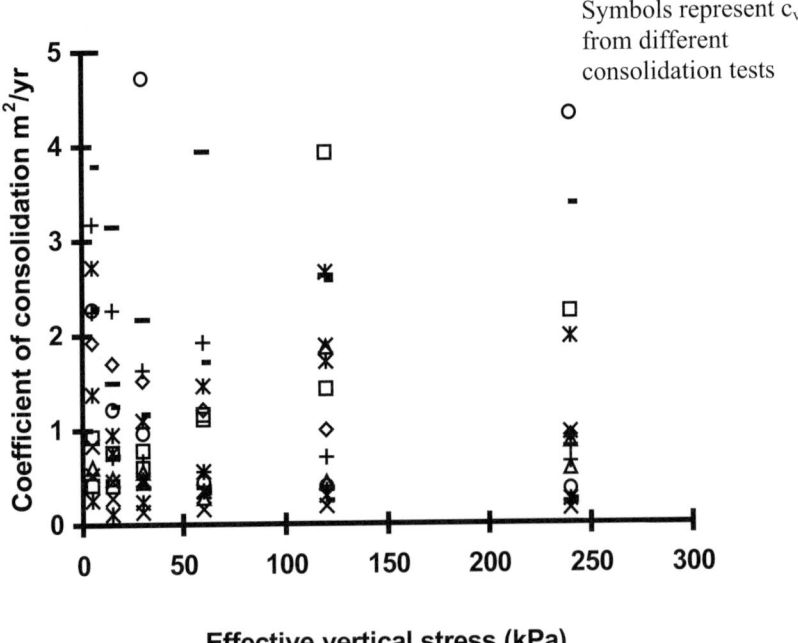

Figure 4 Laboratory determined coefficient of consolidation

Field performance

Initially a 0.5m thick layer of free draining material was laid on the ground to form a drainage layer at the base of the fill which was to be placed on the site. Vertical drains were installed at a triangular spacing of 0.75m and granular fill was placed relatively quickly to bring the site level up to about 6.5m to 7m OD. The rate of filling is shown on Figure 5.

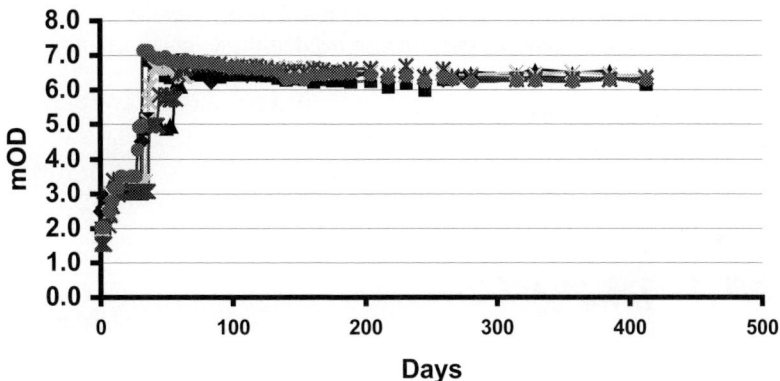

Figure 5 Rate of placement of fill

The field performance was monitored using settlement plates and piezometers. The recorded ground movements at the settlement points are shown on Figure 6.

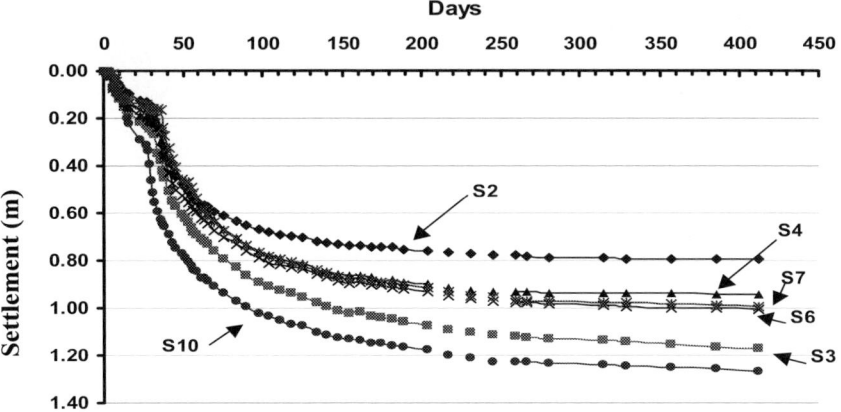

N.B. S2, S3, S4, S6, S7, and S10 indicate settlement points

Figure 6 Recorded movements of settlement plates

The settlements were taken onto steel rods which were connected to a flat steel plate which was placed at the top of the free draining layer (FDM). The ground was made up with side berms to prevent edge shear failure and the six settlement points shown on Figure 6 (S2, 3, 4, 6, 7 and 10) were sufficiently far from the edge of the fill to be considered to be unaffected by edge effects. The variation in the settlements readings is considered to be due to variations in the depth of the soft ground, as well as possible spacial variations in the soil properties.

There are five piezometers (P1, P2, P3, P4 and P5) corresponding to these settlement points and the recorded piezometric head at these locations are given on Figure 7 and Table 1 below.

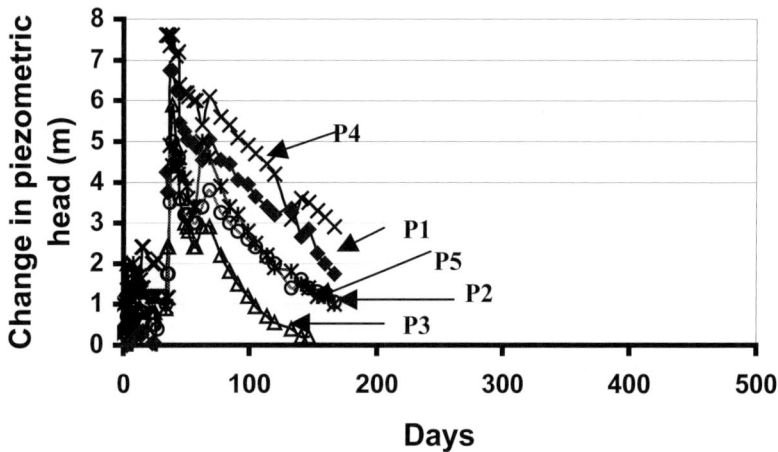

Figure 7 Change in piezometric head

Piezometers	Elevation mOD
P1	-1.502
P2	-1.527
P3	-5.520
P4	-1.430
P5	-2.492

Table 1 Elevations of piezometers prior to fill placement

Analysis of results

The variation in the deposit as shown by the moisture content/depth plot on Figure 1 is typical of these marshland/alluvial soils and case histories from other sites have been reported by Farrell (2000). Most of these soils would be classified as organic Silts or silty Peats. As the soils are normally or lightly overconsolidated, it is useful to relate the variation in the soil properties to the moisture content.

The use of the CPTu, calibrated to the penetration vane, gave a very good indication of the undrained shear strength profile of the ground. In particular it highlighted the upper crustal layer which can be a significant feature in the assessment of the stability of material when placed on these deposits. The value of N_k of 17 to calibrate the CPTu with the *in situ* vanes compares favourably with published data for such soils. As expected from the different stress paths, the triaxial compression tests gave higher c_u/σ_{vc}' ratios (where σ_{vc} is the effective vertical consolidation pressure) than the direct simple shear test (see discussion above). Sufficient fill was not placed to cause an undrained failure, hence no comment can be made on the relationship between the laboratory parameters and the global parameters.

The effective stress parameters of c' of about 0 and ϕ' of 35 to 48° determined in triaxial compression tests are similar to values for these soils on other sites (Farrell, 2000). However, lower values ϕ' are generally recorded in direct shear tests. This is considered to be due to anisotropy, including the arrangement of fibres which would be expected to be predominantly horizontal, where these are present. Thus the difference between the effective stress parameters determined in the various tests can be expected to increase with an increase in organic content.

The piezometers indicated that the total primary settlement was essentially complete in about 250 days, when piezometric pressure readings were adjusted for the settlement of the piezometers themselves. Thus the terminal settlement readings were considered to reflect total primary settlement. The amount of primary settlement predicted by the Asaoka (1978) method from the early readings was also investigated. This method estimates the amount of total primary settlement from recordings of surface settlement movements taken at regular time intervals. This approach necessarily assumes that the entire stratum has a relatively uniform coefficient of consolidation. Figure 8 shows the settlement readings from the site under consideration plotted for time intervals of 25 days with a linear trendline through the points. The total primary settlements computed using the early settlement readings using this approach were between 0.8m and 1.3m which is in line with those finally recorded.

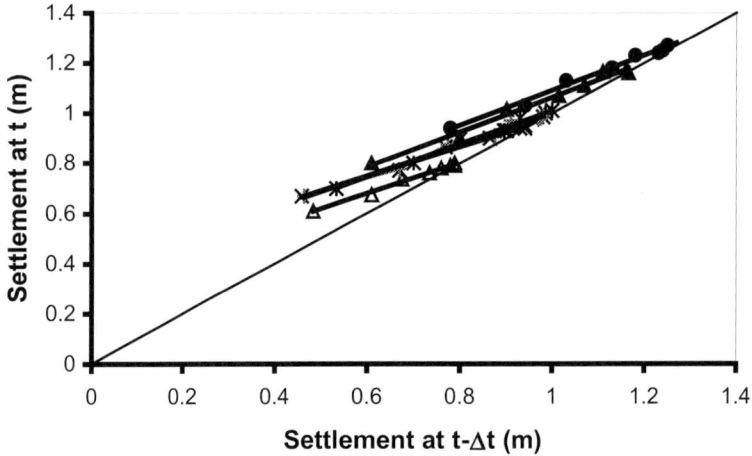

Figure 8 Settlement observations to Asaoka (1978)

The interpretation of the *in situ* compression parameters from the data is uncertain as the distribution of vertical strain with depth was not recorded. Furthermore, it was not possible to get a reliable estimate of the bulk density of the fill because of the presence of large stones. The depth of soft soil at the precise location of the settlement points, as interpreted from the penetration of the vertical drains, varied from about 6m to 7.5m and the actual thickness of fill in the vicinity of the settlement points varied from 5m to 6m. The lower settlement reading was at a point where the depth of soft soil was low and where there was a slight rise in the original ground level which resulted in a lesser thickness of fill than at the other points. It was also necessary to allow for the presence of the upper crustal layer. In the absence of the distribution of vertical strain with depth there is no reliable method of accurately determining the *in situ* parameters of the soil from the data obtained. However it is possible to compare the recorded overall movements with those predicted from the laboratory tests parameters (parameters interpreted from laboratory tests were; $\sigma_c' = 30$kPa, $C_e/(1+e_o) = 0.09$, $C_c/(1+e_o) = 0.3$, $\rho = 1.43$ Mg/m^3) and this comparison is shown on Figure 9.

The computed settlements are based on a bulk unit weight of 20 kN/m^3 for the fill, on the recorded thickness of fill placed, on the depth of soft soil interpreted from the vertical drain installation data and incorporates an allowance for the settlement of part of the fill below the water table. The computed range is within 1m to 1.4m which compares very favourable with the recorded movements listed above which were between 0.8 and 1.3m. However the estimated settlements varied by up to about 40% of values at the actual points. Thus the global behaviour is as predicted but the precise estimation of the settlement at the actual settlement points was less accurate.

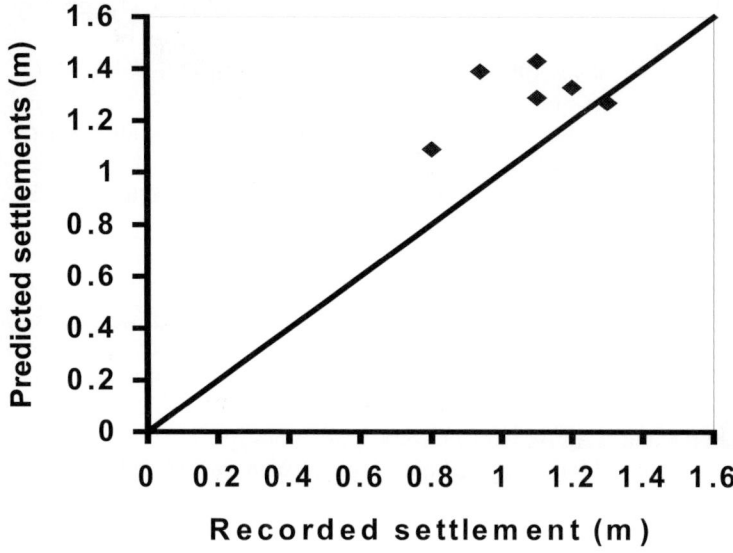

Figure 9 Comparison between estimated and recorded settlements.

The coefficient of consolidation was estimated from the field response using three approaches, namely the piezometric dissipation curves, using the Asaoka's method modified by Magnan and Deroy (1980) and by back analysing a degree of consolidation versus T_r curve from the settlement plate readings.

The coefficient of consolidation was estimated from the piezometric dissipation curves using the method proposed by Nicholson & Jardine (1981) which is:-

$$\delta u_r/\delta t = -(8c_{vh}/(F(n)(2R)^2))u_r \qquad (2)$$

where u_r =   excess pore pressure at radius r
        c_{vh} =   horizontal coefficient of consolidation
        $\delta u_r/\delta t$ = rate of dissipation of excess pore pressure

The values interpreted using this relationship were between 0.5 and 1.8 m^2/yr with an average of 0.9m^2/yr. There are uncertainties when using this approach regarding the actual distance of the piezometer from the drain, particularly with the close drain spacing adopted on this site. A small variation in the verticality of the piezometer or in the vertical drain can significantly affect the coefficient of consolidation estimate using this approach on this site.

Magnan & Deroy (1980) proposed that in the case of combined radial and vertical drainage the coefficient of horizontal consolidation (c_{vh}) is give by

$$c_{vh} = -(\ln(\beta_1)/\Delta t)/((8/((2R)^2 F (n)) + \pi^2(c_{vv}/c_{vh})/(4H_d^2)) \quad (3)$$

Where β_1 is the slope of the straight line on the Asaoka graph; Δt is the time interval used on the plot, d_e is the diameter of the radial drainage cylinder, H_d is the drainage path for vertical drainage, R is the radius of influence of the drain and

$$F(n) = (n^2/(n^2-1))\ln(n)-(3n^2-1)/4n^2 \text{ where } n = 2R/d_e \quad (4)$$

The above formula ignores the effect of smear on the performance of the vertical drains. The equivalent diameter of the vertical drain has been taken as 50mm. The coefficient of consolidation computed using this approach was between 0.8 and 1.2m²/yr with an average of 1.06m²/yr.

Farrell (2000) compared the observed field rate of settlement with that which would be predicted assuming a constant c_{vh} and radial drainage only (see Figure 10). The degree of consolidation (U = settlement at time t/total primary settlement) was related to t/(4R²) which is related to the dimensionless time factor T_r and to the horizontal coefficient of consolidation (c_{vh}) by;

$$T_r = c_{vh}t/4R^2 \quad (5)$$

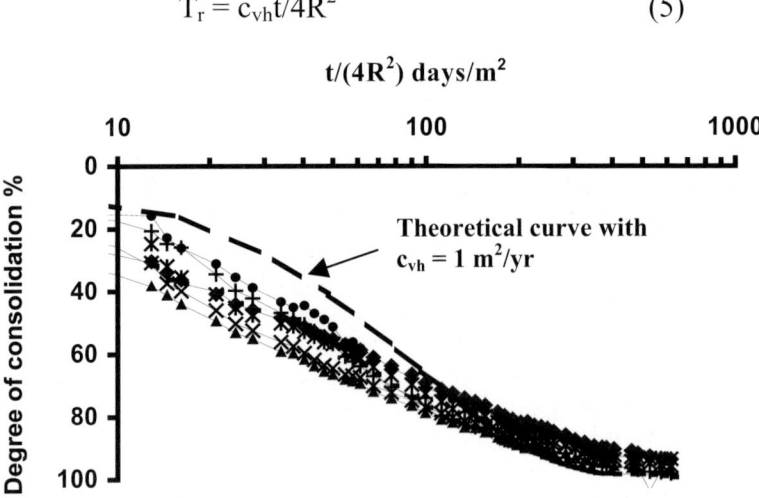

Figure 10 The degree of consolidation versus t/(4R²) for theoretical and observed data

The best fit curve was obtained using a c_{vh} of 1 m^2/yr, however there were indications that the initial rate of consolidation was faster and that it reduced with an increase in effective stress.

The global coefficient of consolidation determined by the three approaches was relatively consistent at about 1 m^2/yr. It must be emphasised that the above approaches do not consider the variation in the coefficient of consolidation with effective stress which can be particularly significant if a stage construction approach were adopted (Farrell, 1997)

Comments and conclusions

There are many recent geological deposits which result in a relatively wide and random variation in soil properties with depth and with aerial extent. This variation adds an additional factor to be considered when estimating the behaviour of compressible soils under ground improvement schemes. The careful analysis of case histories involving construction on these soils gives valuable information on the methods for investigating these soil to obtain the relevant parameters, on the appropriate analysis techniques and on the expected behaviour.

The findings of the case history presented in this paper has shown that there is a good relationship between the undrained strengths interpreted using a CPTu (with an N_k of 17) and those determined from the *in situ* penetration tests using the Geonor H10 vane. The former method, along with simple moisture content versus depth plots, is very useful for getting an overall appreciation of a marshland/alluvial site. The undrained shear strength determined on good quality piston samples in triaxial compression were as expected higher than those determined in direct simple shear tests. The latter is normally considered to represent the average undrained strength mobilised during an embankment failure.

The compressibility of a soil is considered to be best assessed by assuming the normal bi-linear e-$\log\sigma_v'$ relationship and by using a Schmertmann's reconstruction of laboratory results to allow for sample disturbance. This has been found to give good agreement between laboratory and in situ behaviour on such soils (Farrell, 1997). While it is appreciated that the post yield relationship between e and $\log \sigma_v'$ is generally not linear for compressible soils, a reasonable linear approximation can be made for the strain levels normally encountered.

The compressibility of these marshland/alluvial soils is considered to be best expressed as the compression index ratio $C_c/(1+e_o)$. Hobbs (1986) observed that the value of this ratio for peat never exceeded about 0.45. The moisture content of these soft normally consolidated soils are close to their liquid limit. Therefore the moisture contents can be used in place of the liquid limits when the liquid limits are not available. It is therefore possible to compare the value of the compression index ratio for a number of soils. Such a comparison from a number of sites with soils of varying moisture have been combined to form a compression index ratio versus liquid limit plot shown on Figure 11. The results of the case history presented in this paper are agreement with this general relationship.

The amount of primary settlement estimated using the above approach compared very favourable with that recorded in the field. This is similar to the experience from several sites (Farrell, 1997). However this is possibly due to compensating errors as the estimates at particular points from the thickness of soft and the height of fill had greater variation.

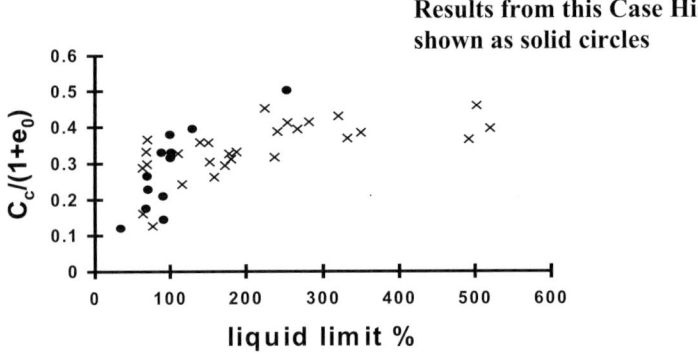

Figure 11 Compression index ratio versus liquid limit

The selection of the coefficient of consolidation is an important aspect of the geotechnical design of ground improvement schemes using preloading and surcharge. There is frequently a difficulty in predicting the appropriate value for design from laboratory test results. The values determined on laboratory samples normally relate to vertical drainage whereas, if vertical drains are used, the horizontal coefficient of consolidation is more relevant. The case history presented in this paper was analysed to back figure the actual global field coefficient of consolidation. Three methods were adopted; (1) the interpretation from the excess pore pressure dissipation curves, (2) the modification to Asaoka's method for consolidation with vertical drains proposed by Magnan and Deroy (1980), and (3) a curve fitting method. All the methods used make the simplified assumption that the coefficient of consolidation is constant over the effective stress range experienced on site, and two of the method assume that the stratum was uniform. The coefficient of consolidation interpreted from the three approaches was relatively close, in the region of 1 m^2/yr. The range of values interpreted from the laboratory tests varied widely and could not readily be related to the field performance. The CPTu dissipation tests recorded values of 4 to 6 m^2/yr, however these reflect effective stresses below the preconsolidation pressure and would normally be reduced by a factor of about 10 to the post yield values.

There were indications in the field readings that the value of the coefficient of consolidation was decreasing at the higher effective stresses and this decrease was not considered in the interpretation methods adopted above. The case history presented in this paper is essentially that the consolidation of the ground under a single load increment. The decrease in the coefficient of consolidation is particularly significant where soft compressible soils are loaded in stages. Morris (1998), back analysed the variation in the coefficient of consolidation of an instrumented stage loaded embankment on alluvial/marshland soils and concluded that the in situ behaviour was similar to that indicated from standard oedometer tests with very low values of the coefficient at the higher effective stress range (*in situ* values of c_{vh} of the order of 0.35 m²/yr were interpreted at the higher effective stresses).

Acknowledgements

The author would like to thank the Barry, O'Connor Gibson and Punch Partnership and Limerick Corporation for permission to publish this article.

References

Asaoka, A. 1978. *Observational procedure of settlement prediction*, Soils and Foundations, 18, (4).

Farrell, E.R. 1997. *Some experiences in the design and performance of roads and road embankments on organic soils and peats*. Proc. of the Conf. on recent advances in Soft Ground Engineering, 1, 66-84, Kuching, Sarawak, Malaysia.

Farrell, E.R. 2000. *Comparison of laboratory predictions versus field performance of embankments over soft ground*, Soft Ground Technology Conf., ASCE Geot. Special Publication No. 112, 435-444.

Hobbs, N. B. 1986. *Mire morphology and the properties and behaviour of some British and foreign peats*, Quarterly Jnl of Eng. Geology, 19, 7-80.

Lunne, T., Robertson, P.K., & Powell, J.J.M. 1997. *Cone penetration testing in geotechnical practice*, Blackie Academic & Professional.

Magnan, J. P. & Deroy, J.M. 1980. *Analyse graphique des tassements observes sous les ouvrages*, Bull. Liaison Labo. P. et Ch., 109.

Nicholson, D.P. & Jardine, R. J. 1981. *Performance of vertical drains at Queensborough By-Pass*, Geotechnique 31, 67-90.

Morris, A. *Behaviour of a basal reinforced embankment on soft organic soil*, PhD Thesis, University of Dublin, Trinity College.

Schmertmann, J. H. 1953. Estimating the true consolidation behaviour of clay from laboratory test results, Proc. ASCE, 79, 1-26.

Settlement or Subsidence: the long-term performance of foundations on peat of the Somerset Levels

E.J. Wilson
E.J. Wilson & Associates, Gloucester GL1 1JJ

Introduction

Peat is an organic soil of generally low shear strength, of very high compressibility, and of extremely high shrinkage potential. Its morphologic and soil mechanics properties are well documented by Hobbs (1986), who also discusses its consolidation properties as have others, e.g. Berry (1983) and Burwash and Weisner (1987). The widely recognised reputation of peat for high compressibility and secondary consolidation has led to its general avoidance as a foundation soil. Indeed its infamy as a foundation material has sometimes led to a predisposition to regard any foundation movement on peat as due to consolidation, whether primary or secondary. Observation of buildings founded on peat has, however, shown that the long-term performance of foundations is dominated by factors unrelated to loading and consolidation settlement. These include the shrinkage due to drainage and desiccation, and the oxidation losses (known as wastage) of dried peat, both of which are widely documented (Schothorst, 1977; Stephens et al, 1984; Waltham, 1989).

Relatively few buildings in England are founded on peat, but the author's investigation of a house on the peat levels in Somerset revealed a characteristic hogging pattern of subsidence (defined as downward flexure of the perimeter or ends of a building: the opposite of sagging) previously observed in the Netherlands and on the Lincolnshire Fens. This inspired a subsequent extension of the investigation by comparison with a number of other houses in the area.

Geology

The Somerset Levels occupy the triangle between Weston-Super-Mare, Taunton and Glastonbury. They consist of a number of broad buried valleys cut into strata of the Trias and Lower Jurassic, and infilled with alluvium of Holocene age up to 15m in thickness, to present a surface level which ranges between 4m and 6m AOD, between east-west ridges of protruding Triassic and Lower

Problematic Soils. Thomas Telford, London, 2001

Jurassic strata. The alluvium consists in its lower part of mainly fine sands and silts, overlain by soft dark grey clays (Kidson and Heyworth 1976). Interspersed are various beds of peat, often lenticular and impersistent, but with a major and relatively persistent bed of peat having its lower boundary on or about Ordnance Datum. Locally this reaches a thickness of 2-3m and more, and its upper surface reaches ground surface over large areas on Tadham, Westhay and Godney Moors to the north of the River Brue, over Chilton Moor, Edington, Calcott, Shapwick, Meare and Glastonbury Heaths to the south of the River Brue, and Kings Sedge Moor and Somerton Moor in the basin of the King's Sedgemoor Drain. The latter peat moor is virtually unpopulated, a few houses stand on Tadham and Westhay Moors, but the area of Shapwick, Meare and Glastonbury Heaths supports a considerable number of houses.

The peat has been extensively dug from its outcrop, by hand as a domestic fuel until the 1960's, then by machine for horticultural use, at an increasing rate. The earlier workings stripped the peat in 200mm layers below topsoil, returning the topsoil for recultivation, until the whole area was lowered, whereupon the cycle was repeated, but the current workings extract the full thickness. This is accompanied by pumped drainage into a system of drainage rhynes.

Case study

The property investigated was a small two-storey house of 100 to 150 years' age, with a number of outbuildings standing in a concrete surfaced yard. The house had hogged along its roofline by an estimated 125mm, and the outer walls leaned outward dramatically. The concrete yard had draped or cambered visibly down towards its edges, and outbuildings leaned radially away from the house, all with sufficient cracking to indicate that this movement was post-constructional. For obvious reasons the individual property is not identified, but it forms reference 20 in Table 1.

The large garden, which extended to some very large willow trees, showed a very distinctive pattern of irregular polygonal cracks, at a spacing of 2m to 5m. These were about 100mm wide, infilled with extremely loose peat debris, and could be probed easily to a depth of almost 1.0m. The owner had recently sunk up to the knee in one of these. The ground surface between the cracks was convex, so that the cracks formed a network of channels some 100mm to 300mm below the lenticular surface between them. These had every appearance of shrinkage cracks. They appeared to be better developed near large trees, and it has been reported that they also occur close to the edge of some of the deeper peat workings.

Area survey

It was apparent that most houses of a similar age in the area showed the same pattern of movement, in contrast to the few modern houses which in general were built on piled or raft foundations, and showed no distortion. In most cases the hogging of the buildings, and any attached outbuildings, was sufficient to be visible from the roadside, with ends visibly "falling off" in extreme cases. It was also noted that this pattern of movement was confined to the outcrop of the peat: those houses founded on alluvial clay showed no visible distortion.

A survey was therefore made in the areas of Tadham Moor, Meare Heath and Glastonbury Heath, by inspecting and photographing buildings frontally from the public road, and by scaling with a straight-edge, to estimate approximately the total magnitude of hogging, in relation to the length, and estimated age, of each building.

A summary of the findings for 10 buildings or building groups on the peat is shown in Table 1.

Ref	Type of Building	Building Age (year)	Estimated Length (m)	Hog* (mm)
1	Rendered cowshed	50	14	100
2	Corrugated clad barn	80	20	140
3	Two-storey house	100	10	100
3A	Single storey ends	80	18	200
4	Two-storey house	100	10	100
5	Two-storey public house	150	14	Irregular
5A	Single storey extension	100	24	200
6	Rendered bungalow	140	8.5	125
6A	Single storey garage ext	60	12	250
7	Single storey pair of houses	50	9	150
14	Rendered bungalow	120	8	125
18	Modern bungalow, with	30	9	Nil
18A	Detached triple garage	30	8	100
20	Two-storey house	140	9	125

* Hog is defined as the maximum difference in level between the centre and the ends of the building

Table 1 Estimated age, length and hog of buildings examined

These figures are no better than indicative, since building age is only guessed from their style, and hogging was crudely measured. They nonetheless show a consistent pattern, approximating in most cases to a hogging of about 1mm but as high as 2mm per year of age, but roughly double that figure for single storey construction and extensions.

Consolidation theory shows that a uniformly loaded area on a flexible foundation will settle more in the centre than at the corners, and each perimeter wall containing such a loaded area would sag. Houses are not uniformly loaded areas, but a uniformly loaded linear foundation should still, if anything, show greater settlement in the centre than at the ends.

The consolidation settlement experienced by a building founded on peat is greater and of longer duration than for one founded on an inorganic soil of comparable shear strength, but the pattern of differential settlement within that structure should be similar, and equilibrium should eventually be reached beyond which no significant further settlement will be experienced. Where an extension is then added to such a building, that extension will undergo a similar history of settlement, but just as on other soils, that edge contiguous with, and partly engaged against the existing building will be partly supported and will settle less, resulting in the extension tilting away from the main building. This pattern is found to greater or lesser degree on all extensions, but on peat it is magnified to an extent which becomes visible, due to the much higher compressibility. It is therefore to be expected that extensions will tilt visibly away from their parent buildings.

What would not be expected, however, is that buildings constructed in a single phase should show the same pattern. The author has, however, observed dwellings on peat in the Netherlands, in Lincolnshire, and now in Somerset, which show this same pattern of long-term settlement.

At least two dwellings constructed on piled foundations some 10 to 30 years previously, were reported to have experienced a gradual lowering of the ground around and beneath them, to the extent that underfloor voids had appeared, large enough in one case to see below the building from one side to the other. Unfortunately in both cases the ground levels had been made up some time before, so that this feature could not be inspected, nor dimensions verified. A hearsay account describes the gaps as two to three inches, i.e. 50mm to 75mm. This was a classic expression of shrinkage and wastage, whereby the ground surface was being lowered without any imposed load, as previously observed (see Waltham, 1989; Stephens *et al.,* 1984).

Dutch research

Niewenhuis and Shokking (1997) reviewed the long-term changes in ground surface levels over wide areas following the drainage of polders, initially to control flooding, with groundwater maintained about 0.1m to 0.2m below surface for some 800 to 1000 years, until the late 1870's when the advent of steam technology increased the pumping rates and hence the freeboard, then more recently to an even greater depth in order to improve trafficability for more mechanised farming. Several interesting conclusions emerged.

1   The long-term subsidence under minimal groundwater lowering amounted to about 2mm/year.

2   The increase in freeboard resulted in an increased rate of subsidence to about 6mm/year, with subsidence being very irregular.

3   Where the peat was covered by even a small thickness of clay, subsidence was about one-third of the amount on exposed peat, and was more evenly distributed.

4   *In situ* settlement gauge measurements showed that the reduction in volume leading to subsidence occurred almost totally within the freeboard zone.

5   Intensive experimentation showed that up to 15% of the subsidence was due to shrinkage on drying, whilst up to 85% was due to the oxidation of organic matter which followed aeration. A small to insignificant proportion was attributable to consolidation settlement due to the increased effective stress due to groundwater lowering.

6   The subsidence rate did not increase significantly for freeboards in excess of about 1m, presumably due to the more restricted access for oxygen.

The term 'wastage' refers strictly to the losses by oxidation, but is often applied to the overall process of volume loss, including the relatively small proportion of pure shrinkage. The term 'wastage' is considered appropriate in this paper as a descriptive term, in the view of confinement of subsidence to the near surface zone. The rates of resultant surface lowering are related to the scale of induced drainage and consequent depth of desiccation, and also to the climate and consequent rate of oxidation (Schothorst, 1977; Waltham, 1989).

Wastage in Somerset

A search of historical Ordnance Survey maps of the Somerset Levels revealed disappointingly few benchmarks and spot heights which could be traced through time to reveal the patterns of regional subsidence. A single benchmark recorded on the 1885 and 1971 surveys indicated 170 mm of subsidence, at a rate of 2.9 mm/year. Of 18 spot heights, half revealed subsidence rates of 2.5-7.6 mm/year. The other half revealed a scatter of results in the range 2.1-14.0 mm/year with the larger error bar being due to conversions from imperial to metric units and small shifts in the spot height locations.

Though the data could not be further refined, they do demonstrate that there is regional subsidence on the Somerset peat, and it is on a scale commensurate with that recorded in the Netherlands and elsewhere.

If the effect of a building or a concrete surfaced area is analogous to that of a small area of clay cover, in that it protects the underlying peat from drying and oxidation, it follows that drying and oxidation will cause a loss of ground generally around, and locally beneath the periphery of each protected area. To use the Dutch figures, the difference between 6mm/year on exposed peat and 2mm/year beneath clay, would give an annual differential subsidence of 4mm/year.

The actual differential subsidence observed in the ten buildings of Table 1 is typically 1 to 2 mm/year, but the area protected by a single house is probably not large enough to offer the degree of protection given by a continuous layer of clay, in that the edge effects of aeration might penetrate beneath the entire building minimising the differential subsidence.

It was also observed that virtually all buildings stand on low platforms raising them above the level of the surrounding ground, by some 300mm to 600mm. While this may reflect, at some sites, a deliberate policy to raise buildings above flood level, the elevation of many is considered to be due to differential subsidence. Exposed ground around buildings subsides by wastage, but the ground beneath buildings is more stable because the peat is at least partially protected from oxidation. In these cases the peat underneath may have been reduced by wastage, while the occupants have periodically reprofiled the ground up to the edge of the building.

Implications for the design of new foundations

The findings have obvious implications for the design of new foundations above peat layers. Whilst the reinforcement to raft foundations must still be designed to spread structural loads, the long-term reality is that the raft will be required to cantilever on all unprotected sides. It is possible that differential subsidence would be minimised by providing a downstand edge, to a depth of about 1.0m, or to the depth of the summer water table if less. These recommendations might seem strange in view of the apparently satisfactory performance of existing rafts. The reality is, however, that rafts not designed to cantilever will be working on an unknown factor of safety.

The traditional alternative is to use piled foundations. Again these appear to perform well. In the long term, however, they provide a rigid foundation surrounded by ground which will decline in level by perhaps 500mm over a 100 year design life. There is therefore an ongoing need to maintain the level of approaches to the buildings, including buried services. Since the loss of ground is close to the surface, though, no significant negative skin friction should develop on the piles.

Implications for buildings insurance

This type of subsidence has obvious significance in terms of insurance cover on existing buildings. Insurance cover for subsidence excludes damage from settlement, and insurers need to be in a position to make a clear distinction. The following seems to be a statement of the most frequently accepted position.

Settlement is the lowering of a structure or part of it due to the compression of ground under the surcharge applied by that structure. All soil and rock materials compress under load: it is only where the magnitude or duration become excessive that problems arise. Normal consolidation settlement diminishes with time, and once complete, any renewed movement for reasons other than an increase in applied structural load, is not settlement and is presumed subsidence.

Subsidence is defined as a lowering of the ground surface, together with any structures thereon, as a result of some post-constructional change in conditions affecting the underlying ground.

Common causes of subsidence include: man-made physical ground loss by mining, tunnelling or deep ground excavation; natural ground loss by internal erosion or solution at depth (e.g. salt, gypsum or limestone), or collapse of infilling to solution features; soil volume changes due to seasonal or other changes in moisture content; soil volume changes due to heat loss to the ground; and changes in effective normal stress due to changes in the groundwater level.

A commonly applied test is that settlement would not have occurred had the structure not been there, whereas subsidence would have occurred irrespective of its presence. This is not infallible, however, since some forms of undeniable subsidence can be triggered by secondary or tertiary consequences of development, e.g. destabilisation of an infilled swallow hole by leakage from drains or soakaways, and heat loss from buildings to the ground. Peat wastage, however, is no exception to this test, since its cause is unconnected with the presence of the building.

Conclusion

The research has identified shrinkage due to desiccation and wastage due to oxidation as the major reasons for subsidence of buildings founded on peat. This explains the apparent anomalous subsidence patterns observed. A consequence of the findings is that foundation movements on peat should be regarded for insurance purposes as subsidence, not settlement.

Acknowledgements

The Author acknowledges permission for the use of information from Cunningham Ellis & Buckle, Chartered Loss Adjusters, on behalf of Prudential Assurance.

References

Berry, P.L 1983. *Application of consolidation theory for peat to the design of a reclamation scheme by preloading.* Quart. Journ. Eng. Geol. 16, 103-112

Burwash, Al & Weisner WR, 1987 *Discussion on "Mire morphology and the properties and behaviour of some British and foreign peats."* by Hobbs, N.B Quart. Journ. Eng. Geol. 20, 97

Hobbs, N.B. 1986. *Mire morphology and the properties and behaviour of some British and foreign peats.* Quart Journ Eng Geol 19, 7-80

Hughes, R.E. 1980. *The use of Ordnance Survey benchmarks for the study of large-scale mining subsidence.* Proceedings of 2nd International Conference on Ground Movements and Structures, (JD Geddes, ed.). Pentech Press

Kidson C & Heyworth A, 1976 *The quaternary deposits of the Somerset Levels.* Quart. Journ. Eng. Geol. 9, 217-235

Niewenhuis, H.S. & Schokking F. 1997. *Land subsidence in drained peat areas of the province of Friesland.* The Netherlands Quart Journ Eng Geol 30, 37-48

Schothorst, C.J. 1977. *Subsidence of low moor peat soils in the Western Netherlands.* Geoderma 17, 265-291

Stephens, J. C., Allen, L. H. and Chen, E., 1984. *Organic soil subsidence.* Geological Society of America Reviews in Engineering Geology, 6, 107-122.

Waltham, A. C., 1989. *Ground subsidence.* Blackie.

Development of methods for identifying problem mudrocks using index tests

J.C. Cripps and M. A. Czerewko
Department of Civil and Structural Engineering, University of Sheffield, S1 3JD

Introduction

Due to their wide distribution, mudrocks are frequently encountered in the course of civil engineering construction work. In addition they may be exploited as a resource for pottery and brick making, as fill material or for the construction of low permeability barriers. They are a diverse group of materials that are distinguished by their dominant grain size of <63 μm. Certain types of mudrocks are susceptible to rapid breakdown to sand and gravel sized fragments in response to changes in stress conditions or moisture content. Mudrocks may host reactive minerals such as pyrite that, on oxidation, give rise to acid and sulfate rich solutions. Besides leading to the degradation of the material, such processes may give rise to environmental pollution and chemical attack of concrete and buried steel structures.

Previous studies have highlighted the importance of clay mineralogy, structure and cementation in controlling physical breakdown and, while it is possible to use these features to identify potentially unstable materials, conventional laboratory assessment is time consuming and expensive to carry out. The alternative approach is to carry out slake durability determinations using the method pioneered by Franklin and Chandra (1972), but experience with this test indicates insensitivity when rating low durability rocks. These problems have been investigated with respect to the evaluation of the mineralogy, texture and structure of a suite of mudrocks of varying durability. Various index tests are appraised in terms of their value for providing predictive tools for identifying mudrocks liable to be susceptible to rapid breakdown due to slaking processes.

Problematic Soils. Thomas Telford, London, 2001

Case histories are described in which physical breakdown and/or pyrite oxidation have led to problems. Due to deficiencies in the current methods for the determination of sulfur compounds in construction materials, reference is made to new procedures for determining sulfur speciation.

Composition and lithological properties of mudrocks

The term 'Mudrock' encompasses claystone, mudstone and siltstone which are generally defined as consisting of >50% siliclastic constituents with >50% being less than 63 μm in grain size (Stow, 1981). Mudrocks are the most widespread of all sedimentary deposits forming 60-70% of the sedimentary sequence. A large proportion of the siliclastic component of mudrocks tends to consist of clay minerals therefore producing their unique engineering properties of plasticity and the propensity to shrink and swell. It has been shown by Shaw and Weaver (1965) that a modal mineral composition of over 400 analysed mudrocks consisted of 60% clay minerals (including detritally derived and diagenetic minerals), 30% quartz, 5% feldspar, 4% carbonates, 1% organic material and <1% iron oxides and traces of mineral phases including sulfide and sulfate minerals such as respectively pyrite and gypsum.

The mineralogy of mudrocks is controlled by many factors including the source and type of clastic input, environment of deposition and diagenetic processes involving low temperature alterations. The clastic input consists of minerals such as quartz, feldspars, organic material, carbonate shell fragments together with phyllosilicate minerals including muscovite mica, chlorite and clay minerals such as illite. In addition, dissolved ions and colloidal phases may enter the depositional environment, these include metalloids such as calcium and magnesium, iron oxides and hydroxides, with dissolved sulphates and carbonates. Depending upon the deposition environment these components may form into various mineral phases that occupy the intergranular pore space. Carbonates including calcite form under marine conditions where as siderite (iron carbonate) forms under partly reducing and brackish conditions. Under reducing marine and non-marine conditions within the shallow sediment surface or under anoxic bottom water conditions pyrite may form where the input of detrital iron, iron oxide, iron hydroxide and sulfur from dissolved sulphates or organic sources are present in sufficient quantities.

Upon burial, diagenetic transformations of certain clay mineral constituents take place within mudrocks. These occur in response to an increase in the overburden pressure related to the thickness of the overlying sediment and temperature increases. Such diagenetic modifications bring about changes in mineralogy and structure of the sediment. A number of associated processes such as progressive illitization of smectite occur. At depths below approximately 500m deep the removal of expandable layers from clay structures and decomposition of kaolinite to chlorite occurs. Reactions involving the conversion of these less stable minerals into more stable clay minerals are accompanied by the release of Si^{4+} and Ca^{2+}, which results in the precipitation of quartz and carbonate cements. Burial pressures also cause clay minerals to develop strong diagenetic bonds that bind them together.

Effects of diagenesis and burial on mudrock properties

Although the mineralogical changes referred to above are important, the main physical post-depositional process affecting mudrocks is compaction (Rieke and Chilingarian, 1974). Progressive burial causes the fabric of clays and mudrocks to be in a state of continual change. Initially, depending upon the electrolytic condition, the clays are deposited as aggregates referred to as flocs with a open cardhouse or honeycomb type structure or as individual grains (Van Olphen, 1963; Moon and Hurst, 1984). In these two depositional forms, clay particles, which are platy in shape, are formed into edge-to-edge or face-to-face arrangements depending on the electrolyte concentration and cation valency. This gives the resulting deposit a relatively high porosity, in the region of 70 to 90%, and a high water content. Compaction associated with deep burial eventually destroys such structures, so the resultant clays develop a tighter packing. The porosity of a clay deposit decreases rapidly for the first 300-500m of burial (Baldwin, 1971). This change is governed by a number of factors which include composition, rate of deposition, pore structure and permeability, availability of permeable zones to allow the removal of pore water, decomposition of organic matter, chemical diagenetic processes and the state of the interstitial fluids (Chilingarian, 1983). The process of burial compaction converts the clay sediment into rock by means of lithification, and as such the physical properties of the resultant material progressively change. Constituent grains are brought into closer contact as water is expelled from pore spaces. If the non-clay clastic constituents are brought into contact a framework of pores is provided and clay minerals tend to be squeezed and re-orientated to fill these voids. Where no clastic framework develops the reduction in porosity is associated with reorientation of clay particles so that the crystallographic Z-axis tends towards a vertical orientation.

During formation and early diagenesis swelling clay mineral species such as smectite and mixed layer illite-smectite are commonly present in mudrocks. As these mudrocks tend to be poorly indurated and have a high tendency to disaggregate upon immersion in water, they are referred to as compaction mudrocks (Mead, 1936). The process of disaggregation is referred to as slaking and is a unique property especially in mudrocks and weathered rocks containing clay mineral species, it is defined as the disintegration of material subjected to alternate cycles of wetting and drying. As already mentioned, with the resultant conditions due to progressive burial the less stable reactive clay mineral species become converted to more stable clay mineral species, for instance illitization of smectite and mixed layer clays and conversion of kaolinite to chlorite or illite and precipitation of carbonate and silica cements. These indurating mineralogical changes are accompanied by physical changes including micro-discontinuity closure, mineral realignment, reduction in void space and porosity and an increase in density. These latter changes, which characterise cementation mudrocks (Mead, 1936) also enhance durability by restricting access of air and water to the any remaining reactive phases and they therefore are beneficial in terms of the engineering performance of mudrocks.

These textural changes mentioned play an important role in changes to the weathering performance and engineering behaviour of the resulting mudrocks. The strong clay bonds formed during burial have the ability to release strain energy on a time-dependent basis once the rocks undergo prograde processes such as uplift and weathering (Taylor, 1988; Dick and Shakoor, 1992; Dick, 1992).

Durability of mudrocks

In terms of durability, mudrocks may possess desirable or undesirable engineering properties depending upon their degree of compaction, structure and mineralogy. Such behaviour, which is a function of wetting and drying under atmospheric conditions, takes two forms, physical and physiochemical (Taylor, 1988). In less indurated or non-cemented mudrocks, it has been found that a few cycles of wetting and drying caused rapid breakdown. This is attributed by Taylor (1988) to air breakage within voids and along discontinuities together with negative pore water pressure formation on drying which leads to tensile failure of weak inter-crystalline bonds (Kennard *et al.*, 1967). Russell (1982) in his study of Ordovician mudrocks, discovered that the degree of micro-fracturing represents an important role in the rate of breakdown of mudrocks, as during wetting and drying, breakdown is initiated along the fractures by air breakage due to capillary suction of water (Badger *et al.*, 1956). Certain species of clay minerals such as smectite have been attributed to giving mudrocks their swelling and slaking tendencies, but Taylor & Smith (1986) point out that many British mudrocks of Carboniferous and older age do not contain discrete smectite phases and yet these properties still persist. They attribute the factor controlling expansion to be the cation types present in clay minerals. However, it is clear that voids including pores and microfractures also govern swelling, as they provide the means by which water gains access to clay minerals (Dick and Shakoor, 1992). Morgenstern & Eigenbrod (1974) found that there is a direct correlation between strength loss during weathering of mudrocks and changes to the initial void ratio, bulk density and initial moisture content. It therefore can be seen that the engineering properties of swelling, slaking and strength in mudrocks are controlled to a large extent by physical factors such as porosity, void ratio, dry density and moisture content as well as the bulk mineralogy, clay mineral presence and type.

The slaking properties of mudrocks were of concern to the British coal mining industry, which established the 'Shale Panel' in 1953 (Taylor, 1988). Their research established a basic understanding of abrasive and breakdown behaviour of mudrock (Badger *et al*, 1956). The Shale Panel designed the 'end-over-end' slaking test, which was the forerunner of the ISRM slake durability test (Franklin and Chandra, 1972), which is the accepted standard test for durability classification of mudrocks. Subsequently much research has been carried out on mudrocks using these and other test procedures and in summary the research projects have shown that the principal processes involved in mudrock degradation are:

Cause	Physical feature
Stress relief upon exhumation-	Discontinuities (micro-fractures, voids)
Degree of induration (maturity)-	Slaking (due to tensile failure as a result of air & water access and/or swelling of clay minerals present)
Diagenetic history-	Swelling (as a result of clay minerals present)
Mineralogy-	Chemical weathering (pyrite oxidation)

Table 1 Principal processes in mudrock degradation.

It has been shown by Czerewko & Cripps (2001) that the standard slake durability test lacks sensitivity when it is used to distinguish between durable and non-durable mudrocks. This is due to the inherent bias possessed by the equipment used for the test, as an arbitrary 2mm mesh size is used to distinguish between slaked and non-slaked material. Fragments of mudrock used for testing may disintegrate considerably although depending on the size of the slaked debris they do not actually pass through the test drum adding to the lost material. A solution suggested in the standard is to sieve the slaked material but this procedure is seldom adopted, due to time and cost. Although the one cycle test generally proved adequate for characterising very durable mudrocks, it was found that further breakdown of the majority of samples tested occurred beyond the first cycle. Although some breakdown beyond the third cycle occurred in some samples, the three cycle test provided a more reliable assessment of durability for a wide range of mudrocks and for certain samples including cemented mudrocks a five cycle tests was found to be more useful for classification.

The ISRM test (1979) which is dynamic in nature tends to be very aggressive such that complete breakdown of low-durability materials occurs, this may give an inaccurate evaluation for mudrocks which on exposure such as in cuttings may only deteriorate at a very slow rate. It is therefore suggested that the modified jar slake test, used either on its own or in conjunction with the slake durability test, provides a more accurate and convenient method for evaluation of the slaking behaviour of mudrocks spanning a wide range of durabilities.

Investigation of laboratory testing procedures

Mudrocks can be characterised in terms of mineralogy, chemistry and physical properties. Although a mineralogical and chemical investigation provides knowledge of the mineral suite present it cannot in itself be used to predict the engineering behaviour of the rock. Mudrocks are found in various stages of induration, where the principal variables are consolidation and cementation. The latter are a reflection of burial and diagenetic history of the material. Therefore it was decided to carry out detailed characterisation of the selected suite of 49 mudrock samples including routine analytical and physical tests and also novel or modified test procedures. The characterisation and testing is divided into four broad categories: (1)-rock classification, (2)-textural analysis, (3)-mineralogical investigation and (4)-bulk physical and engineering characterisation. These are described briefly here but further details are presented in Czerewko, (1997).

Mudrock classification

A detailed classification of the mudrock was carried out based on descriptive and analytical results including the following. A detailed classification of the rock type based on the clastic content of the sample (percentage of quartz and feldspars) as determined by X-ray diffraction (Spears, 1980 and Taylor, 1988). A detailed description of structural features such as laminae and the extent of fissility development, based on hand specimen descriptions using a binocular microscope, and where possible textural description using back-scattered scanning electron microscopy and standard petrographic thin section analysis. Characterisation of the diagenetic rank of the material based on illite 'crystallinity' X-ray diffraction measurement and vitrinite reflectance work by which the progressive increase in illite formation and maturation and changes to organic matter are measured.

Textural analysis

A detailed textural classification was carried out on each sample using a new procedure based on back-scattered scanning electron microscopy and the diagenetic rank parameter classification. Fracture distribution counts were carried out on sections cut perpendicular and parallel to bedding. It was found that the modified jar slake index (Czerewko & Cripps, 2001) provides an additional means of textural classification since the test monitors the extent of microfracture development.

Mineralogy

A detailed investigation into the mineralogical make up of each sample was undertaken. This consisted of X-ray diffraction analysis of the whole rock fraction to determine the mineral suite present in each sample and also quantification of quartz and feldspar minerals. The clay mineralogy was also determined and quantified based on X-ray diffraction analysis of the <2 μm size fraction. The samples were also characterised geochemically using X-ray fluorescence analysis from which mineral phases such as titanium minerals were quantified. Additionally wet chemical procedures were used for the determination of quartz, carbonate minerals, sulfate and sulfide species and organic carbon.

The quantification and characterisation of the quartz and feldspar present, clay mineralogy, especially the quantity of mixed layer clays or smectite present, and cement phases, including carbonates, and organic carbon content which has been found effectively to improve the durability of mudrocks were essential aspects of the assessment of the durability. Minor potential reactive phases, especially pyrite, have a tendency to result in swelling and deterioration of the rock and surrounding materials due to oxidation, dissolution resulting form the acidic conditions generated, and conversion to other mineral phases. In addition, the methylene blue adsorption test was investigated. The test is a method of determining the cation exchange capacity of clays therefore determining a mudrocks potential reactivity resulting from its clay mineral composition. It was intended to see how well the test correlated with engineering behaviour. Fairburn and Robertson (1956) showed that there was a remarkably good positive correlation between the methylene blue index value and the Atterberg limits of soils. It has also been found that the methylene blue index is a simple, reproducible means of characterising the clay mineralogy in a whole rock sample (Taylor,1967, Stapel & Verhoef, 1989 and Cokca & Birand, 1993).

Bulk physical and engineering characterisation

A detailed investigation into the bulk physical characteristics and engineering properties of the samples was undertaken using standard and novel procedures.

Basic physical characteristics which were determined include moisture conditions such as natural moisture content, moisture absorption and moisture adsorption, specific gravity determined on powdered material, dry density and porosity determinations including total porosity and relative porosity.

Engineering parameters include slaking potential, swelling potential and strength were determined. The slaking potential was determined using the Franklin slake durability test which was performed over five cycles (Brown, 1981; Taylor, 1988) where a value of Id_3 >60% is used to distinguish between non-durable and durable mudrocks. The slaking potential was also determined using the modified jar slake test which unlike the slake durability test is a non-dynamic test (Czerewko and Cripps, 2001). The ISRM (Brown, 1981) triaxial unconfined swelling cell was used for measuring the strain on all three axes during water immersion induced swelling of dried cube shaped specimens of the mudrocks. In addition, the Gibbs & Holtz (1956), powder free swell test was also used, this test procedure eliminates any structural controls on the swelling measuring only the influence of mineralogy on swelling. Strength determinations on mudrocks are difficult to carry out and the most effective means of determining strength of mudrocks is the point load test (ISRM, 1985) since it requires a minimum of sample preparation. This test procedure was used to characterise all the samples. In addition, where material was able to withstand the process of coring conventional Unconfined Compressive Strength tests

(UCS) and direct and Brazilian disc tensile strength determinations were undertaken.

Evaluation of mudrock durability based on simple index tests

Once a detailed analysis had been completed for the mudrocks suite parametric studies were carried out on the data and it was found that strong correlations existed between a number of the index tests and their mineralogical, textural and engineering properties. From the correlation analysis for the mudrock lithotypes (Czerewko, 1997), the controls presented in Table 2 were found to govern the mudrocks durability which are presented with the relevant index tests to evaluate these controls. Of the index tests evaluated (Czerewko, 1997), the jar slake index, moisture absorption and methylene blue adsorption are considered to offer the highest level of confidence in a meaningful characterisation of mudrocks, the tests are easy to perform and the results reproducible.

Mudrock type	Durability controlling features	Index test
Argillite	Strength and durability mainly controlled by structural features.	Moisture absorption, jar slake index, microfracture index, methylene blue index
Claystone	Mineralogy (expansive clays and organic carbon), discontinuities.	Loss on ignition, moisture adsorption, moisture absorption, methylene blue index.
Mudstone	Textural features such as microfractures.	Jar slake index, moisture absorption.
Mudstone-fissile & laminated.	Textural features and mineralogy	Jar slake index, moisture absorption, moisture adsorption, microfracture index, methylene blue index.
Siltstone	Textural features (such as porosity and microfractures) and clay mineralogy.	Jar slake index, moisture absorption, microfracture index, methylene blue index.

Table 2 Control features in mudrock types and relevant index tests.

It is rarely possible nor affordable to carry out a detailed characterisation of mudrock samples as was carried out in the study. Although it can be seen from Table 1 that certain index tests may be used for the general characterisation of mudrock samples and the identification of potentially problematic material which would require further testing. A predictive matrix approach based upon simple index tests can be applied to characterisation of mudrocks by means of a ranking system (Czerewko & Cripps, 1998). This allows selection of the control parameters which are considered to be applicable to a particular situation. For example, an index characterization matrix based on the jar slake index

(Czerewko and Cripps, 2001), moisture absorption (ASTM C97, 1987 and Czerewko, 1997) and methylene blue index (ASTM C837, 1986; Stapel and Verhoef, 1989; Czerewko, 1997), which are ideal index tests for characterizing all mudrock lithotypes could be weighted as follows-

1. Jar slake index. Determines the slake durability of the samples, see Table 3.

Ij	Classification	Rank [A]
1 - 3	Extremely durable	1
3 - 6	Moderately durable	2
6 - 8	Non durable	3

Table 3 Rank values for jar slake classification of mudrocks.

2. Moisture absorption. Determines the textural maturity of the sample, see Table 4.

Moisture absorption %	Diagenetic rank parameter	Rank [B]
> 6	<6.5	3
3 - 6	7 - 9.5	2
<3	10 - 12	1

Table 4 Rank values for moisture absorption versus DRP.

3. Methylene blue index. Which is indicative of the potentially reactive mixed-layer clay content of the sample. A figure of 22% mixed-layer clay content was taken as the critical content after Taylor and Spears (1970), who found that effective swelling and slaking problems were commonly encountered in samples of >22% mixed layer clay. The ranking values are presented in Table 5.

MBA	%MLC	Rank [C]
<1	0	1
1 - 2.3	0 - 22	2
>2.3	>22	3

Table 5 Ranking values of MBA versus %MLC.

The results from the selected index tests when summed produce a rank durability value for each sample which may be used for assessing a range of samples, these values should be then compared with the results from selected laboratory characterization tests as a means of standardisation. A rank durability values for 41 of the samples tested in the study by Czerewko, (1997) were calculated and compared against a durability classification based on laboratory determined values of Slake durability (Id_3) and Volumetric swelling (Ev%), these are presented in Table 6 for 38 of the samples. Rank values of 1 to 3 are

Sample	Id$_3$	Ev%	RANK			Rank Total Value	Classification
			A	B	C		
Ca11	99.0	0.005	1	1	1	3	Extremely durable
Ca12	99.4	0.015	1	1	1	3	Extremely durable
Ca13	99.4	0.029	1	1	1	3	Extremely durable
O11	56.9	2.07	3	2	3	8	Non-durable
O21	98.5	0.03	1	1	1	3	Extremely durable
O31	98.4	0.165	1	1	1	3	Extremely durable
S11	94.6	1.08	2	2	2	6	Durable
S31	98.6	0.43	1	1	1	3	Extremely durable
D11	94.9	0.46	2	2	1	5	Durable
D12	97.6	1.17	2	2	2	6	Durable
D21	98.4	0.19	1	1	1	3	Extremely durable
C1B1	97.5	0.875	2	1	2	5	Durable
C1B2	91.8	1.69	2	1	2	5	Durable
C1B5	85.0	1.65	3	1	2	6	Durable
C21	46.8	3.12	3	3	3	9	Non-durable
C31	97.5	0.63	1	1	2	4	Extremely durable
C41	74.3	1.15	3	3	3	9	Non-durable
C51	27.8	2.79	3	3	2	8	Non-durable
C52	44.8	0.285	1	1	2	4	Non-durable
C61	56.2	1.49	3	2	2	7	Non-durable
C71	85.2	2.55	3	2	2	7	Non-durable
C82	22.3	3.39	3	3	3	9	Non-durable
C83	81.7	0.32	1	2	2	5	Durable
C92	98.0	0.22	1	1	2	4	Extremely durable
C101	63.2	1.86	3	2	3	8	Non-durable
C111	56.9	6.3	3	3	3	9	Non-durable
C112	81.2	0.28	1	2	2	5	Durable
C121	75.2	3.79	3	2	2	7	Non-durable
C122	85.6	0.7	2	1	2	5	Durable
C131	41.4	2.18	3	2	2	7	Non-durable
C132	91.0	0.61	2	2	2	6	Durable
C133	10.4	1.53	3	2	2	7	Non-durable
C141	63.3	2.92	3	3	3	9	Non-durable
C151	72.4	0.83	2	2	2	6	Durable
C161	93.8	0.26	2	3	2	7	Non-durable
C171	91.1	0.84	2	2	2	6	Durable
C172	94.8	0.33	2	1	2	5	Durable
C181	90.7	0.87	2	2	2	6	Durable

A = Jar slake index : B = Moisture absorption : C = Methylene blue index.

Table 6 Classification of 38 mudrocks based on the rank durability approach.

classed as extremely durable samples which were not prone to swell or slake, values of 4 to 6 are classed as durable samples, and values of 7 to 9 are classed as non-durable samples. It can be seen from the results that the method is capable of identifying all the potentially problematic samples in the data set. For the durable samples, results indicate material that could experience potential loss in durability, these samples show a rank value of 6, and would require further detailed testing.

Detrimental effects associated with mudrocks as a result of chemical weathering of pyrite

In recent years problems have been encountered in various construction projects including embankment dam construction (Davies & Reid, 1997 etc), building foundations (Hawkins & Pinches, 1987, Cripps & Edwards, 1997 etc), highway schemes (Thaumasite Expert Group, 1999), and tunnelling (Bracegirdle et al., 1996) which have been attributed to the presence of sulfates and sulfides in the ground material. The occurrence of sulfur compounds has resulted in the corrosion of buried steel structures and also problems due to thaumasite sulfate attack on buried concrete members.

In geological material sulfur compounds commonly occur as sulfides and sulfates. Sulfides such as pyrite form under anoxic marine and non-marine conditions in which many mudrocks are formed and therefore they are a common constituent of mudrocks. Once mudrocks containing pyrite are exposed such as in mining or civil engineering works, the free access of air and water into the material results in rapid oxidation of pyrite which produces acidic, sulfate rich conditions. The acidic sulfate rich waters will react with calcite, if present, precipitating gypsum, which may result in floor heave to structures. The sulfate rich water is also associated with ettringite and thaumasite attack on buried concrete structures. Sulfates such as gypsum are also commonly found in mudrock deposits containing evaporite sequences such as Mercia Mudstone and also sulfates may occur in small quantities in Coal Measures mudrocks. Therefore if oxidation of sulfide species associated with sulfate species occurs much higher concentrations of sulfate in solution will result due to the much higher concentration of sulfates in acidic waters. This can lead to corrosion of construction materials and the production of leachate, which can pollute surface waters and groundwater.

As a result of the problems caused by sulfur species to highways structures, research commissioned by the Highways Agency (Reid *et al.*, 2001) has highlighted the need for adequate testing of ground material in association with highways works. The research found that existing testing protocols are inadequate for the determination of sulfur species and has proposed a new four stage testing procedure for classification of materials based on (1)-Water soluble sulfates, (2)-Acid soluble sulfates and monosulfide, (3)-Total reduced sulfides, and (4)-Total sulfur determinations.

Conclusions

In mudrock material durability is the most important engineering property. Durability is affected by the slaking and swelling properties and strength which are dependant on the physical and mineralogical properties of the rock. The mineralogical and physical properties are themselves a result of the geological and diagenetic history to which the mudrock has been subjected. Therefore a detailed characterisation of mudrock samples requires a systematic approach and may involve the use of necessary and yet costly analytical techniques such as X-ray diffraction, scanning electron microscopy and mercury porosimetry. Where many samples are to be analysed such as in a major construction project which will have suitable funds allocated for laboratory testing, a representative selection of samples should be subjected to detailed laboratory characterisation of the mineralogy, texture and engineering properties. In the study of Czerewko (1997) it was evident that certain simple index tests correlate strongly with fundamental aspects of mudrock mineralogy and texture, which control their engineering behaviour. Therefore it was suggested that for efficient mudrock characterisation, a ranking system based on selected simple index tests is used for routine characterisation and detailed laboratory investigations on selected representative samples. This procedure is also useful in assisting detailed characterisation of limited samples in situations where detailed laboratory testing would not be viable. In addition to the routine physical tests and limited mineralogical testing which is routinely used in mudrock investigations, testing should be undertaken to determine the presence and quantity of sulfide mineral species. Sulfide minerals, in particular pyrite, are commonly present in mudrocks, and on exposure to atmospheric conditions are liable to oxidise producing acidic conditions detrimental to the environment and also to steel and concrete structures. In addition oxidation of sulfide minerals may precipitate sulfate minerals which can cause expansion with detrimental effects on engineered structures such as foundations.

It can therefore be seen that with the use of simple index tests and specific testing to determine the sulfide content, a mudrock may be characterised suitably such that any engineering or environmental problems that may be liable to occur will have been considered.

References

A.S.T.M, C97. 1987. *American Society for Testing and Materials, Soils and Rock; Building Stones:* annual book of ASTM standards 4.08, ASTM, Philadelphia, Pennsylvania, 1189.

A.S.T.M, C837. 1986. *American Society for Testing and Materials, Standard test method for methylene blue adsorption index of clay.* ASTM, Philadelphia, Pennsylvania, 15.02, 275-276.

Badger, C.W.,Cummings, A.D & Whitmore, P.L. 1956. *The Disintegration of Shales in Water.* Journal of the Institute of Fuel, 29, 417-423.

Baldwin, B. 1971. *Ways of deciphering compacted sediments.* J. Sediment. Petrol., 41, 293-301.

Bracegirdle, A., Jefferis. S.A., Tedd. P., Crammond. N.J., Chudleigh. I. & Burgess. N. 1996. *The investigation of acid generationwithin the Woolwich and Reading Beds at Old Street and its effect on tunnel linings.* In: Geotechnical Aspects of Underground Construction in Soft Ground (Mair & Taylor, ed), Balkema, 653- 658.

Brown, E.T. 1981. *Rock characterisation, Testing and Monitoring.* Pergamon, Oxford.

Chilingarian, G.V. 1983. *Compactional diagenesis.*In: Sediment Diagenesis (Eds. A. Parker and B.W. Sellwood). Reidel Pub. Co., Dordrecht, 57-168.

Cokca, E. & Birand. A. 1993. *Determination of cation exchange capacity of clayey soils by the methylene blue index test.* Geotechnical Testing Journal, GTJODJ, 16, 518 - 524.

Cripps, J.C. and Edwards. R.L. 1997. *Some geotechnical problems associated with pyrite bearing mudrocks.* In: Ground Chemistry Implications for Construction. A.B. Hawkins (ed). 77-86.

Czerewko, A. 1997. *Diagenesis of Mudrocks, illite crystallinity and the engineering properties of mudrocks.* Unpublished PhD thesis. University of Sheffield.

Czerewko, M.A. & Cripps. J.C. 1998. *Simple index tests for assessing the durability properties of mudrocks.* In: The Geotechnics of Hard Soils-Soft rocks, Evangelista & Picarelli (eds). Balkema, Rotterdam.

Czerewko, M.A. & CRIPPS. J.C. 2001. *Assessing the durability of mudrocks using the modified jar slake index test.* Quarterly Journal of Engineering Geology and Hydrogeology, 34, 153-163.

Davies, S.E. & Reid. J.M. 1997. *Roadford Dam: Geochemical aspects of construction of a low grade rockfill embankment.* In: Ground Chemistry Implications for Construction. A.B. Hawkins (ed). 111-131.

Dick, J.C. *1992. Relationships between durability and lithological characteristics of mudrocks.* Unpublished PhD thesis, Kent State University, 235.

Dick, J.C. & Shakoor. A. 1992. *Lithological controls of mudrock durability.* Quarterly Journal of Engineering Geology, 25, 31-46.

Fairburn, P. E. & Robertson. R. H. S. 1956. *Liquid limit and dye adsorption.* Min. Soc London Clay Mins. Bull., 3, 129 - 136.

Franklin, J.A. & Chandra. A. 1972. *The slake durability test.* International Journal of Rock Mechanics and Mineral Science, 9, 325-341.

Gibbs, H.J. & Holtz, W.G. 1956. Engineering properties of expansive clays. Trans. Am. Soc. Civ. Eng., ASCE, 121, 641-663.

Hawkins, A.B. & Pinches. G.M. 1987. *Sulfate analysis on black mudstones.* Geotechnique, 37, 191-196.

I.S.R.M, 1985. *Suggested method for determining point load strength.* International Society of Rock Mechanics Commission on Testing Methods. Int. J. Rock Mech. Mining Science, 22, 51 - 60.

I.S.R.M, 1979. *Suggested methods for determining water content, porosity, density, absorption and related properties, and swelling, and slake-durability index properties.* International Journal of Rock Mechanics and Mining Science and Geomechanical Abstracts. 16, 141-156.

Kennard, M.F., Knill. J.L. & Vaughan. P.R. 1967. *The geotechnical properties and behaviour of Carboniferous shale at the Balderhead Dam.* Q. J. eng. Geol. London, 1, 3-24.

Mead, W.J. 1936. *Engineering geology of dam sites.* Transactions of the 2^{nd}International Congress on Large Dams, Washington, DC, 4, 183-198.

Moon, C.F & Hurst. C.W. 1984. *Fabric of muds and shales: an overview. In: Fine-grained sediments: deep-water processes and facies.* (Eds. D.A.V. Stow and D.J.W. Piper). Blackwell Scientific Publications, Oxford, 579-593.

Morgenstern, N.R. & Eigenbrod K.D. 1974. Classification of argillaceous Soils and Rocks. *Proceedings of the American Society of Civil Engineers, Journal of the Geotechnical Division*, 100, 1137-1156.

Reid, J.M., Czerwko. M., Cripps. J.C. & Hiller. D.D. 2001. Sulfate specification for structural backfills. TRL Project Report PR/CE/163/99.

Rieke, H.H. & Chilingarian. G.V. 1974. *Compaction of Argillaceous Sediments. Developments in Sedimentology*, 16, Elsevier, 424.

Russell, D.J. 1982. *Controls on shale durability: the response of two Ordovician shales in the slake durability test.* Canadian Geotechnical Journal, 19, 1-13.

Shaw, D.B. & Weaver. C.E. 1965. *The mineralogical composition of shales.* J. Sediment. Petrol., 35, 213-222.

Spears, D.A. 1980. *Towards a classification of shales.* Journal of the Geological Society, London, 137, 125-129.

Stapel, E.E. & Verhoef. P.N.W. 1989. *The use of the methylene blue adsorption test in assessing the quality of basaltic tuff rock aggregate.* Engineering Geology, 26, 233-246.

Stow, D.A.V. 1981. *Fine-grained sediments:Terminology.* Q. J. eng. Geol. London, 14, 243-244.

Taylor, R.K. *Methylene blue adsorption by fine grained sediments.* J. Sediment. Petrol., 37, pp. 1221 - 1230. 1967

Taylor, R.K. 1988. *Coal Measures mudrocks: composition, classification and weathering processes.* Quarterly Journal of Engineering Geology, London, 21, 85-99.

Taylor, R.K. & Smith. T.J. 1986. *The Engineering Geology of Clay Minerals: Swelling, Shrinking and Mudrock Breakdown.* Clay Minerals, 21, 235-260.

Taylor, R.K. & Spears. D.A. 1970. *The breakdown of British Coal Measure rocks.* International Journal of Rock Mechanics and Mineral Science, 7, 481-501.

Thaumasite Expert Group. 1999. *The thaumasite form of sulfate attack: Risks, diagnosis, remedial works and guidance on new constructions.* Department of the Environment, Transport and Regions, London.

Van Olphen, H. 1963. *An introduction to Clay Colloid Chemistry.* Interscience, 301.

Building on landfill – a sustainable solution for the 21st Century

G.B. Card[1] **and D.J. Richards**[2]
1) Card Geotechnics Limited, Aldershot, Hants. GU11 1TY
2) Department of Civil and Environmental Engineering,
   University of Southampton, Southampton, Hants, SO17 1BY

Introduction

Planning and design of development on landfill should aim to reduce the risks from landfill gas to a safe level. A problem arises determining the degree of risk and what constitutes a safe tolerable level. This paper reviews current planning issues of development on landfill and the use of risk management framework techniques to allow building development affected by landfill gas. It also discusses recent design techniques and describes scientific research currently being undertaken on waste degradation and the rate and magnitude of secondary processes. The research is aimed at providing sustainable building solutions and improving understanding of landfill gas generation and migration, as well as settlement processes.

Development and planning framework

Development controls

Decisions on development of closed or completed landfill sites are matters for the relevant local planning authority. The, Planning and Compensation Act (Do E, 1991) introduced the duty for planning authorities to determine applications for planning permission in accordance with the Development Plan unless material considerations consider otherwise. Environment Agency Draft Technical Report CWM 172/98 (EA 1998) provides technical guidance on assessing the completion of licensed landfill sites and identifies the issues to be considered to determine whether or not a landfill is likely to cause environmental pollution or harm to human health. These issues are also identified in Department of the Environment Circulars 21/87 (DoE 1984) and 17/89 (DoE 1989) that outline the measures needed to conduct a survey of a completed landfill to assess its suitability for development. ICRCL Guidance

Problematic Soils. Thomas Telford, London, 2001

Note 17/78 (1990) should also be consulted with regard to development issues on landfill and contaminated land. More recent advice to planning authorities on the development of contaminated land, including landfill, is contained within national policy guidance documents, in particular PPG 10 (DETR, 1999) and PPG 23 (DETR, 1999).

Development on landfill

In line with government policy to develop brownfield sites a significant proportion of new development in the UK is being built on gassing ground, typically in former industrial or urban areas, as well as docklands, reclaimed estuarine areas and on or adjacent to landfill sites.

The Government's guidance on tackling the issues of contaminated land is set out in the DETR Circular "Environment Protection Act 1990: Part IIA Contaminated Land" (DETR, 1990) which came into force on 1 April 2000. The Circular reinforces the principle that a site should be assessed on the basis of, the now familiar, "source, pathway and receptor" concept for assessing the risk from contamination. On the basis that the risk actually exists the remediation adopted is based on the 'suitable for use' approach with due regard to likely current and proposed future activities of the land. This approach is increasingly being adopted to evaluate the suitability of development on or adjacent to landfill.

Building development is at greatest risk from landfill gas ingress from the ground where:

- large volumes of degradable material exist
- large volumes of gas are generated creating a positive driving pressure and advective flow from ground to beneath or within a building
- also where small volumes of gas are being generated but the development offers direct pathways to sensitive receptor locations, e.g. the use of stone columns in gassing ground to support building foundations.

In the case of operational or closed landfills with large volumes of potentially bio-degradable material, landfill gas can be generated rapidly in large volume with a high surface emission rate. The majority of incidents connected with landfill gas are from such sites and they pose a higher risk to building development.

Risk management framework

General considerations

Planning and design of any development should aim to reduce the risks from landfill gas to a safe acceptable level. A problem arises determining the degree of risk and what constitutes a safe acceptable level. The benefit and quality of living standards provided by the provision of the motorcar and services such as gas and electricity in the home gives rise to the public acceptance of their inherent danger and risk. Fatal incidents involving motorcars or domestic gas explosions have not deterred the public opinion from accepting and using such appliances. General advice on the use of risk assessment and acceptable risk criteria for land-use planning is given in HSE (1989 and 1994) and DoE (1995) and these can be used in landfill risk management. Specific guidance on the assessment of risk for land affected by landfill gas can be found in O'Riordan and Milloy (1995).

Public opinion and attitudes change and although technical problems relating to gas protection can generally be overcome, other considerations such as legislation, economics, commercial marketability, maintenance of control measures and local authority requirements may preclude certain forms of development and/or gas protection measures. Barry et al, (1999) provide useful examples of the use of risk assessment and risk management criteria for development planning of closed landfills.

Evaluation techniques

Two approaches can be used for gas protection design based on assessment of risk to development:-

1. A site specific quantitative approach considering frequency and distribution of gas concentrations, borehole flow rate and estimated surface emission rate and the nature of the source of generation. Risk assessment may be used to assess optimum combination and adequacy of gas protection measures to specific type of development and land-use.

2. The use of generic tables or charts, such as Table 28 from CIRIA Report 149 (Card, 1994) or DETR (1997). This approach is becoming more frequently used, however, these tables should not be used as a definitive design tool.

The use of generic tables and charts can often lead to a wide inconsistency in gas protection requirements for sites with similar gas regimes, and frequently over-conservatism in the design (Wood and Griffiths, 1994). Harries *et al.* (1995) identify that interpretation of gas regimes requires an understanding of the gas generation process and the means of gas escape from the ground. Therefore the most important aspect of relating the gas regime below a site to the risks it poses to any development is the surface emission rate, i.e. how quickly the gas is coming out of the ground. The lower the surface emission rate the lower the risk.

Gas protection systems

Risk based design
The primary consideration for development on landfill is to judge whether the conceptual development layout proposed is suitable for its intended purpose based on a thorough knowledge and understanding of the ground conditions and the gas regime. Conceptual design of buildings/structures should address three important factors in relation to gas protection:

- the design and construction should itself provide a barrier to gas migration and ingress
- the design and construction should encourage dilution and dispersion of gas if, for whatever reason, it should accumulate beneath or within the building/structure
- the design should be environmentally sustainable in line with current good practice set out in BS7750 (1996) and guidance documents such as CIRIA Report 149 (Card, 1994).

In most situations, managed engineering solutions can be found to provide gas protection measures. However, certain types of measures may not be aesthetically acceptable or would be inappropriate to the nature of the development. For example, the use of in-ground active abstraction and the erection of vent or flare stacks within a residential development would jeopardise the commercial marketability of the properties, quite apart from difficulties in maintenance of the abstraction system. In these circumstances, better use of the land might be made by less sensitive end-use development in which the gas protection measures are acceptable to the general public, building occupants and owners.

For all new development it is necessary to comply with Building Regulations -Approved Document Part C, (DoE, 1992) to ensure integrity, fitness for occupation and standards for protection against groundwater ingress, damp-proofing, thermal insulation, natural ventilation and fire precautions should also be met. In addition, from April 1999 the Building Regulations – Approved Document Part M (DoE, 1999) requires all new building development to have facilities for disabled access. This will include the provision of either level access from ground level to inside floor level or ramped approaches. Features necessary to comply with new regulations can conflict with the need to provide gas protection measures to buildings, such as underfloor passive ventilation, external wall vents etc. and careful design detailing is required to avoid conflict. Basic considerations for building gas protection are shown in Table 1. Useful guidance on detailing gas protection measures for buildings is provided in BRE Report 414 (Johnson, 2001).

Barriers to gas migration and ingress	Dilution and dispersion of gas
Use of low permeability building materials and products, e.g. solid rather than cavity block walls, mass concrete and low permeability membranes	Minimise downstanding beams, crosswalls etc., beneath ground slabs that provide confined spaces for gas accumulation.
Use of low permeability construction techniques, e.g. in situ mass concrete with steel reinforcement to limit shrinkage cracking.	Provide sloping underside of ground slab to encourage spillage and dispersion to atmosphere of gas.
Avoiding construction that provides joints/openings for gas entry, e.g. pot and beam ground slabs.	Maximising underslab ventilation by decreasing volume of undercroft and increasing number of vents.
Use of precast suspended or composite floor construction or raft construction to provide complete oversite barrier beneath structure	Detailing interior rooms for maximum ventilation and movement within acceptable limits.
Minimising or eliminating where practicable penetrations through the ground slab, particularly service entries	

Table 1 Basic considerations for gas protection of buildings (after Card, 1994)

Selection criteria

An alternative approach for new development is to remove/reduce the gassing regime prior to development using in-ground venting techniques to dilute and disperse gas. In this way the overall scope and level of gas protection measures required to buildings and structures can be reduced. A decision flow chart for evaluating development suitability and the appropriate form of gas protection measures is shown in Figure 1.

For certain types of development, particularly such as service chambers/rooms and ducts, etc., it may not be possible to provide and maintain a safe atmosphere because of the confined nature of the sub-structure and limited ventilation. In these circumstances the conceptual design of the substructure should incorporate a layout so that as far as possible routine access is not required to areas where hazardous gas is likely to be present or could accumulate. In contrast sub-structures such as large basement car parks are usually designed to be waterproof as well as require ventilation to disperse vehicle exhaust emissions. These design requirements can by default provide protection from ingress of landfill gas from the surrounding ground.

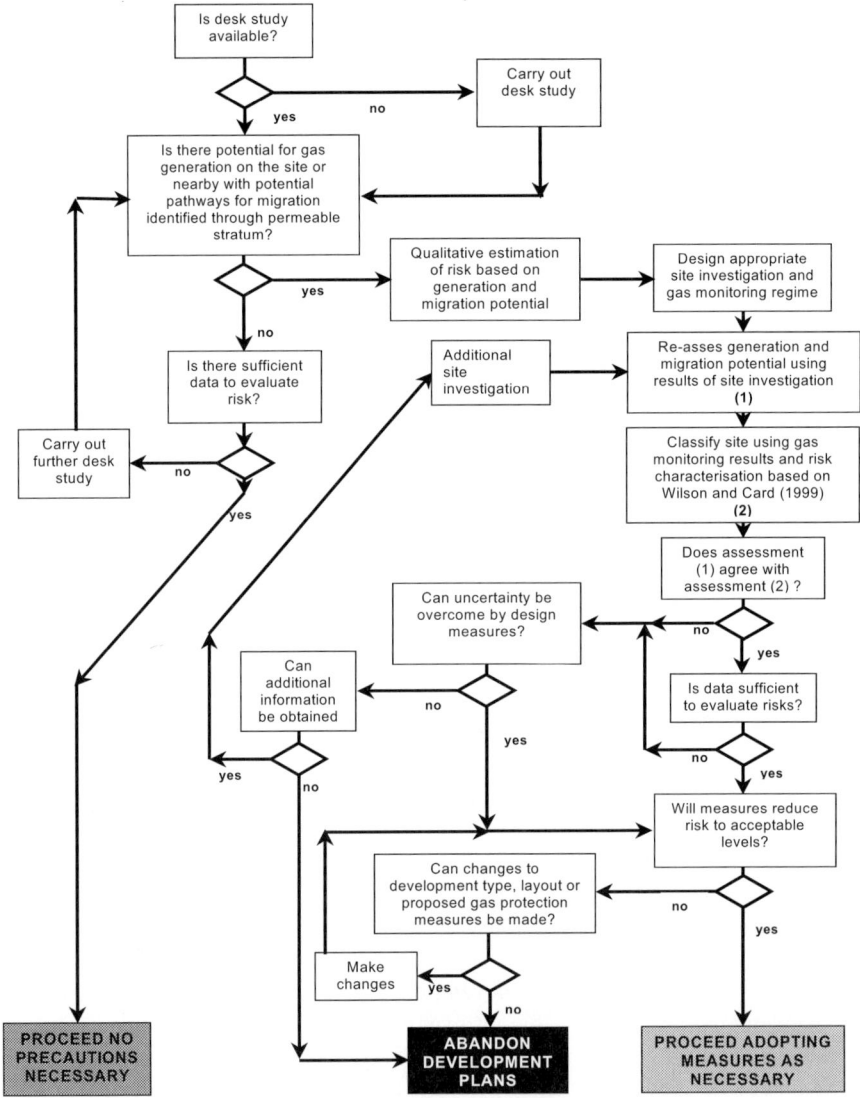

Figure 1 Flow chart for evaluating gas protection measures and type of development.

New techniques

New building techniques and materials are constantly being introduced to the industry that can offer gas protection. One example is described by the SCI (2000) and comprises new composite steel and concrete modular construction techniques for floors and foundations to provide a gas resistant structure for building development. A key design element of this form of construction has been to embrace many of the requirements described in Table 1 for gas protection.

Another example is the development of new insitu gas venting techniques. Wilson and Shuttleworth (2001) describe a new system for constructing a combined passive venting barrier using geocomposite materials.

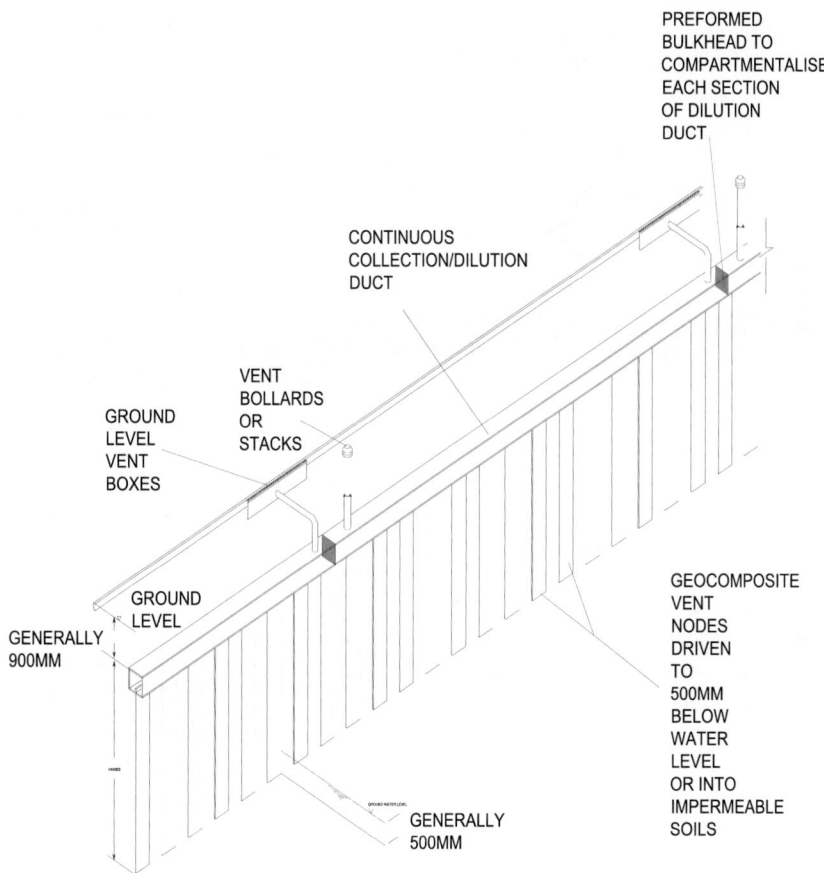

PREFORMED BULKHEAD TO COMPARTMENTALISE EACH SECTION OF DILUTION DUCT

CONTINUOUS COLLECTION/DILUTION DUCT

VENT BOLLARDS OR STACKS

GROUND LEVEL VENT BOXES

GROUND LEVEL

GENERALLY 900MM

GEOCOMPOSITE VENT NODES DRIVEN TO 500MM BELOW WATER LEVEL OR INTO IMPERMEABLE SOILS

GENERALLY 500MM

Figure 2 A schematic layout of a passive dilution barrier

The concept of the passive dilution barrier is to form a low pressure area relative to the surrounding gassing ground, to encourage gas to flow towards the barrier. This is achieved by driving discrete vent nodes into the ground, which are connected to a collection/dilution duct running along the top of the strips. The nodes comprise highly efficient geocomposite strips. The duct has a high flow of fresh air through it by means of passive ventilation. This is one of the key advantages of the system as it:

- dilutes gas emissions to tolerable levels,
- causes a venturi effect in the geocomposite vents, which enhances gas flow from the ground towards the vents.

Ventilation of the duct can be achieved using a combination of vent stacks, bollards or ground level boxes, depending on the gas regime and wind conditions at a particular site. A schematic layout for the barrier is shown in Figure 2.

Scientific advances

The risk based selection of gas protection measures and their design requires a detailed understanding of the chemical and physical behavior of landfill gas together with its generation and migration characteristics. Increasingly, research is focusing on the study of secondary settlement of landfill where the process of gas generation is a key aspect of the overall performance of the materials.

Mechanisms of waste settlement are normally classified as either primary or secondary (Powrie *et al.*, 1998). Primary settlement, due principally to mechanical compression of the waste mass, is caused by crushing, distortion, reorientation, bending and/or breaking of waste particles as vertical stresses are increased, either during compaction or due to the self weight of the fill as further material is deposited. Primary settlement does not therefore affect the post-closure behaviour of a landfill particularly in respect of gas generation.

Secondary settlement of waste results primarily from biological decomposition of the organic fraction and physio-chemical processes such as corrosion and oxidation, together with an accompanying mechanical rearrangement of the particles known as ravelling. While there are data concerning the secondary settlement of landfills (e.g. Bjarngard and Edgers, 1990; Bleiker *et al*, 1995; Kostantinos *et al*, 1997; Watts and Charles, 1999), these have generally been used to produce empirical models of waste settlement using a secondary consolidation (creep) based approach. Such models were originally developed within a conventional soil mechanics framework to describe purely mechanical phenomena in saturated or partially saturated soils. Consequently, they do not address the key factors influencing secondary settlement in a waste landfill; are therefore unable to account for changes in these key factors from case to case; and cannot be used as predictive assessment tools encompassing the whole range of physio-chemical processes and physical changes that typically take place within the waste mass.

Quantitative investigations into the fundamental mechanisms of secondary settlement are rare. One of the main difficulties is the range of factors influencing the processes that cause secondary settlement, which include:

- the composition (e.g. degradable/reactive fraction) of the waste,
- the as-placed density and depth,
- waste pre-treatment (e.g. shredding, baling, composting or partial initial degradation),
- active gas extraction,
- water content, and
- leachate mobility and quality.

High quality large-scale laboratory studies into these factors are prerequisite for the development of the quantitative predictive methods needed to assess rates and magnitudes of secondary processes in existing landfills. Such studies should also aim to provide a clear framework for determining the implications for post-closure landfill management, for example, from implementation of the EC Landfill Directive.

To assess secondary settlement effects linking these factors (i.e. waste composition, settlement and gas potential together with gas and leachate mobility and quality), a series of closely controlled laboratory tests on $0.16m^3$ samples of waste is currently being undertaken within the Waste Management Research Group at the University of Southampton, as part of the Engineering & Physical Sciences Research Council Waste and Pollution Management Programme. The samples are held under a constant applied stress (up to the equivalent of 40m of landfill overburden) within purpose-built large-scale cells contained within a constant load consolidation machine. Leachate is constantly recirculated through the waste to accelerate gas generation rates.

Non-destructive chemical analyses of the leachate and measurements of the degree of settlement of the waste are also undertaken. The chemical analyses allow overall carbon and nitrogen balances to be carried out as well as providing details of the progression of the degradation process. The analytical programme includes the measurement of:

- dissolved inorganic and organic carbon and nitrogen (DIC\DOC and DN) in the leachate using a carbon/nitrogen analyser;
- total carbon and nitrogen (TC and TN) of the waste;
- volatile fatty acids (VFAs) by GC/FID;
- ammonia (derived from the breakdown of proteinaceous material);
- chemical oxygen demand (COD);
- gas generation rate and composition by GC/FID/TC;
- heavy and transition metal analyses after microwave digestion of solid and leachate samples followed by AAS and/or HPLC/UV detection.

The degradability of the wastes tested in the laboratory is further quantified by means of BMP testing in small scale laboratory cells. These smaller test cells are dismantled and sacrificed at different stages of the degradation process, with the whole sample used for destructive chemical testing.

Field data for the project is being supplied by Building Research Establishment (BRE) Centre for Ground Engineering and Remediation and Golder Associates (UK) Ltd from site monitoring projects they have undertaken and directed as part of the high level of industrial support that this research has attracted. These data together with the results of the large and small scale laboratory tests are assisting in the development of quantitative predictive tools for the assessment of secondary compression in landfills, taking due account of initial conditions such as the composition and density of the waste, and operational conditions such as leachate and gas management protocols. These tools are based on landfill behaviour and process models that are currently being developed at Southampton and at Napier University.

The structure of the Southampton model is based on that proposed by El-Fadel *et al.* (1996), and by Young (1989) who described a mathematical model of the methanogenic ecosystem present in degrading wastes. The chemistry of the degradation reaction proposed by Young (1989) is used in conjunction with the catalytic role of biomass described by El-Fadel *et al.* (1996) to control the rate of reaction. The El-Fadel *et al.*, (1996) model is based on a carbon balance and therefore does not explicitly track the water content of the waste, which is an important factor in modelling waste consolidation. This is remedied by the introduction of the equations given by Young (1989). A description of the model, which also incorporates an empirical relationship between dry density, hydraulic conductivity and effective stress derived from data from the large scale compression cell at Pitsea (Beaven & Powrie, 1995; Beaven, 1999) can be found in White *et al.*, 2001.

The Napier model, which is currently being funded by ESART under the Landfill Tax Credit Scheme, adopts a coupled hydraulic, biochemical and mechanical approach to the analysis of landfill settlement. This approach allows for a more fundamental interpretation of the key factors affecting secondary settlement. For example, the coupled hydraulic and biodegradation models enable leachate quality/movement and the progress of biodegradation to be simulated. The impact of different site management practices such as leachate recirculation can also be investigated.

There is very little guidance available for designers and regulators to assess the impact of secondary waste processes on, for example cap performance (Powrie *et al.*, 1998), and/or the post-closure maintenance and aftercare requirements of landfills. This poses a serious threat to the long-term integrity of landfill sites and their viability for future building development, with potentially damaging consequences for the environment. The research work described above, forms a small part of the high quality research currently being undertaken in the UK to further understand the behaviour of landfill material and to develop alternative methods of disposal. This research is therefore timely due to an increasing general awareness of the potentially damaging impact of landfill waste disposal. Operators, regulators and designers need to be able to predict with confidence the likely consequences of secondary processes within landfill, in order to develop adequate long-term protection measures.

Conclusions

The current development planning framework in England and Wales together with recent policy guidance presented in Part IIa of the Environment Protection Act 1990 (DETR, 1990) allows the use of risk management techniques to evaluate site selection and the potential for development on landfill. The same techniques can also be used to select and design sophisticated gas protection systems which rely on a knowledge of landfill gas behavior. For these systems to be technically robust, however, demands a greater understanding and quantification of the behavior of landfill gas, in particular its generation potential and migration characteristics. Current research into these issues together with numerical modeling techniques will provide a better understanding and prediction of long-term gas generation and surface emission rates, as well as develop an understanding of the influence of degradation processes after long-term settlement.

References

Barry, D.L., Summersgill, I. M., Gregory, R.G. & Hellawell, E.E. 1999. *Remedial engineering for closed landfill sites.* CIRIA Report CP/61, Construction Industry Research and Information Association.

Beaven, R.P., 1999. *The hydrogeological and geotechnical properties of household waste in relation to sustainable landfilling.* PhD dissertation, University of London.

Beaven, R.P. & Powrie, W. 1995. *Determination of hydrogeological and geotechnical properties of refuse using a large compression cell.* Proceedings of the 5th International conference on landfills, Sardinia 95 (T H Christensen, R Cossu and R Stegmann, eds.) 2, pp 745-760. Cagliari: CISA Environmental Sanitary Engineering Centre.

Bjarngard, A. & Edgers, L. 1990. *Settlement of municipal solid waste landfills. Proceedings of the 13th Annual Madison Waste Conference*, 192-205.

Bleiker, D.E., Farquhar, G. & McBean, E. 1995. *Landfill settlement and the impact on site capacity & refuse hydraulic conductivity.* Waste Management & Research 13, 533-554.

BS7750, 1996. *Specification for Environmental Management Systems.* British Standards Institute.

Card, G.B. 1994. *Protecting development from methane.* CIRIA Report 149, Construction Industry Research and Information Association.

DoE, 1987. Planning Circular 21/87. *Development of contaminated land.* Department of the Environment.

DoE, 1989. Planning Circular 17/89. *Landfill sites: development control.* Department of the Environment.

DoE, 1991. The Planning and Compensation Act, Department of the Environment.

DoE, 1992. The Building Regulations 1991. Approved Document C, *Site preparation and resistance to moisture.* Department of the Environment and The Welsh Office, HMSO, 1992 Edition.

DoE, 1995. *A guide to risk assessment and risk management for environmental protection.* Department of the Environment. HMSO, 92pp., ISBN 0 11 743091 3.

DoE, 1999. The Building Regulations 1991. Approved Document M, *Access and facilities for disabled people.* Department of the Environment and The Welsh Office, HMSO, 1999.

DETR, 1990. Environment Protection Act 1990, Part IIA, Contaminated Land Department of the Environment Transport and Regions.

DETR, 1997. *Passive venting of soil gases beneath buildings, Guide for Design.* Vol. 1. Department of the Environment Transport and the Regions, Partners in Technology, prepared by Ove Arup and partners, September 1997.

DETR, 1999. Planning policy guidance note 10: *Planning and waste management.* PPG10. Department of the Environment Transport and the Regions.

DETR, 1999. Planning policy guidance note 23: *Planning and waste management.* PPG23. Department of the Environment Transport and the Regions.

EA, 1998. Environment Agency, Draft Technical Report CWM 172/98

El-Fadel, M., Findikakis, A.N. & Leckie, J.O. 1996 *Numerical modeling of generation and transport of gas and heat in landfills.* Waste Management and Research 14, 483-504.

Harries, C.R., Witherington, P. J. & McEntee, J. M. 1995. *Interpreting measurements of gas in the ground.* CIRIA Report 151, Construction Industry Research and Information Association.

HSE, 1989. *Risk criteria for land-use planning in the vicinity of major industrial hazards*: A discussion document. London, HMSO.

HSE, 1994. *Quantified risk assessment*: *Its input to decision making.* London, HMSO.

ICRCL, 1990. *Notes on the development and after-use of landfill sites.* ICRCL Guidance Note 17/78 Eighth Edition. December 1990 Interdepartmental Committee on the Redevelopment of Contaminated Land

Johnson, R W. 2001. *Protective measures for housing on gas-contaminated land.* Building Research Establishment Report BR 414.

Kostantinos, A, Papachristou, E & Georgios, D. 1997. *Settlement measurements at the msw landfill of Thessaloniki, Greece.* Proceedings of the 6th International conference on landfills, Sardinia 97 (T H Christensen, R Cossu and R Stegmann, eds.) 3, 545 - 549. Cagliari: CISA Environmental Sanitary Engineering Centre.

O'Riordan, N. J. & Milloy, C..J. 1995. *Risk assessment for methane and other gases from the ground.* CIRIA Report 152, Construction Industry Research and Information Association.

Powrie, W., Richards, D.J. & Beaven, R P. 1998. *Compression of waste and implications for Design.* Symp. Geotechnical Engineering of Landfills (N. Dixon, E.J. Murray and D.RV. Jones, eds), 3-18, EMGG and ICE, Thomas Telford.

SCI, 2001. *Innovation in composite ground floors and piling for housing.* Steel Construction Institute, Report RT175, DETR Partners in Technology (*In print*)

Watts, K.S. & Charles, J. A. 1999. *Settlement characteristics of landfill wastes.* Proceedings of the Institution of Civil Engineers (Geotechnical Engineering) 137, October, 225-233

White, J. K., Robinson, J. & Ren, Q. 2001. *A framework to contain a spatially distributed model of the degradation of solid waste in landfills.* Proceedings of the 8th International conference on landfills, Sardinia 2001 (T. H. Christensen, R. Cossu and R. Stegmann, eds) Cagliari: CISA Environmental Sanitary Engineering Centre. In Press.

Wilson, S.A. & Shuttleworth, A. 2001. *Design and performance of a passive dilution gas migration barrier.* Ground Engineering (*in press*)

Wood, A.A., & Griffiths, C. M. 1994. *Debate: contaminated sites are being over-engineered.* Proceedings of the Institution of Civil Engineers. August 1994, 102, 97-105

Young, A. 1989. *Mathematical modeling of the methanogenic ecosystem,* J. Chem. Tech. Biotechnol.

Reclamation of contaminated land with specific reference to Pride Park; Derby

D. Bunce and P. Braithwaite
ARUP Environmental, Solihull, West Midlands, B90 8AE

Introduction

Pride Park is an 80 ha former industrial site on the edge of Derby City Centre which was previously used as a domestic and industrial landfill, gas and coke works, rail engineering plant and for gravel extraction. These uses have left behind a cocktail of contamination including high levels of oils, tars, phenols, heavy metals, ammonia, boron and low level radioactive waste.

In 1992, Derby City were successful in winning a bid for £37.5M of City Challenge Funding from the Department of Environment with the aim of investing in the inner city areas of Derby to provide development and employment opportunities. Pride Park was the flagship development and some £20M of funding was used to develop the derelict land into a mixed commercial, leisure and residential development. Ove Arup & Partners (Arup) were appointed as reclamation Engineers in 1993 by Derby City Council and have subsequently developed and implemented a reclamation strategy for the site.

It was found that the site was heavily contaminated as a result of its previous land uses and in order to prevent migration of contaminants, a 3 km long bentonite containment wall was constructed around the most heavily contaminated part of the site. The containment wall created a basin in which the water levels were expected to rise as a result of rainwater falling on the site and inflows from beneath the wall and the underlying aquifer. Thus a £2.3 million groundwater abstraction system was installed consisting of 18 abstraction wells located around the site from which the water continues to be pumped to the Groundwater Treatment Plant before it is discharged into the River Derwent. The plant was built on a modular basis to allow for future modifications should the nature of contamination in the groundwater change in the long term. A 40,000 m^3 fully engineered waste repository was also constructed on site within the bentonite containment wall to safely encapsulate the worst contaminated matter.

Contamination

Previous land use on the Pride Park site included Litchurch gas and coke works, a domestic/industrial landfill on the eastern half of the site and a locomotive works on the western half. In addition some of the site to the north was used for gravel extraction. These land uses left the site heavily contaminated with oils, tars, phenol, heavy metals, asbestos, ammonia, boron and some low level radioactive materials.

The initial site investigation works included records for 800 soil samples. Each of these samples was tested for 22 different contaminants ranging from oils and tars to heavy metals. These soil test results were divided into groups of metals, inorganic compounds and other hazards, giving five sets of contaminants: toxic metals, phytotoxic metals, cyanides, organics and miscellaneous.

The first challenge was to logically appraise a significant quantity of contamination data from the site. In order to do this a site specific classification system was developed based on the guidelines available in 1993 including those from the ICRCL (Interdepartmental Committee for the Redevelopment of Contaminated Land), the Kelly guidelines and the Canadian Council of Ministers of the Environment (CCME). The classification system identified contamination levels on a scale of 1 to 4, with 1 as relatively uncontaminated and 4 as the worst level of contamination. Groundwater samples were also classified on a 1 to 4 scale derived from EC Directives.

Detailed models of the contamination on the site were developed to assist in the reclamation design. These models indicated that the site could generally be split into eastern and western halves. The former comprised of old landfill and gas works sites with the worst contaminated soils and even more highly contaminated ground water. The latter including the gravel pits and engineering works had localised and impersistent areas of contamination with generally uncontaminated ground water.

Reclamation strategy

The two principal objectives of the reclamation strategy issued in November 1993, were to minimise off-site disposal of the contaminated soils and to ensure that contaminants do not migrate into the River Derwent (adjacent to the site on the north and east boundaries).

Developing the strategy was simplified by the conditions found on site. To the east the high contamination levels extended to 10m below the surface, which made treatment impractical and uncommercial. The area has thus been safely contained within a 600mm minimum width bentonite cement vertical cut-off wall, sealed 1m into the underlying Mercia Mudstone. A high density polyethylene (HDPE) membrane was placed centrally within the wall to ensure that the design permeability of 10^{-9} m/s was achieved.

Construction of the 3km long wall was complicated by the number of existing services crossing the site, which involved constructing the wall around the services whilst maintaining the wall's integrity. At the surface a 5m wide, 500mm thick clay cover to prevent the bentonite cement drying and cracking capped the wall. A gas-venting trench is to be constructed to encircle the closed landfill to prevent pressure build-up or landfill gases migrating to neighbouring sites.

Parts of the landfill and old gas works sites will be surfaced with permeable capillary break blanket as individual developments require. Their 650mm thickness and the grading of the stone material are designed to ensure that in periods of drought the capillary rise of any contaminant will be less that the blankets' thickness. Rain can percolate through but end users will be protected. As rain passes through the waste, soluble contaminants will be carried to the base of the fill as leachate.

To minimise the amount of materials to be removed from site, a purpose-built, fully engineered landraise, is to be constructed within the bentonite cut-off wall to take 40 000m³ of class 4 soils from the rest of the site. Its design includes a composite clay/HDPE lining with leachate drainage and landfill gas wells, and it will have an impermeable cap, over which a soil covering will permit vegetation growth and landscaping. The western part of the site has required less intensive treatment, with local removal of contaminated soils.

Water modelling

A two-stage hydrogeological analysis was undertaken to model groundwater conditions before and after construction of the cut-off wall. The first stage programme AQUA set the model boundary conditions as either 'no-flow' or 'fixed-head', and predicted the natural groundwater conditions within the site. The parameters were calculated using the known groundwater abstraction at the adjacent sand and gravel quarry within the modelled area and comparing the predicted groundwater drawdown with that measured in the field.

Prior to the construction of the bentonite containment wall extensive hydrogeological modelling of the groundwater was carried out to predict the effect that the wall would have on both internal and external water levels. A localised increase of 0.7m in the groundwater level outside the wall was predicted as a result of impedance of the ground water flow created by the wall.

Field pumping trials and further hydrogeological analysis were used to predict the effects on the groundwater levels within the wall. The main finding of this analysis, was that the construction of the wall would in effect create a basin in which water levels would be expected to rise. The water rise is attributable to the following sources (see Figure 1):

- rainwater falling on the site and percolating through the permeable capillary break blanket

- water flowing into the contained area under the containment wall due to differential heads inside and outside the wall

- water flowing up from the underlying Sherwood Sandstone aquifer

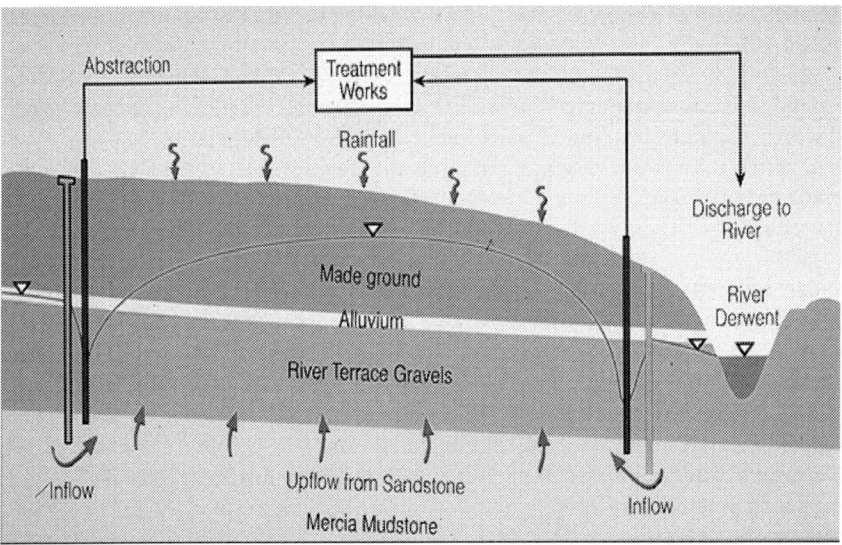

Figure 1 Schematic illustration of the various hydrogeological aspects at the site

The second stage of hydrogeological analysis was to predict the effect of the wall on water levels within it as well as outside. It commenced with field pumping trials in six boreholes in the gravels and comprised step and constant rate tests. As the situation was far from ideal, in that the aquifer is finite in area, varies in thickness, is heterogeneous, and anisotropic, constant rate data for each borehole was analysed using three separate techniques: 'confined', 'leaky' and 'unconfined'. Distance-drawdown calculations used parameters from the pumping trial data to establish drawdown at the wall as a result of pumping from abstraction boreholes.

Data from the pumping trials was incorporated into the stage 2 AQUA modelling with the range of permeabilities deduced from the pumping trial data assigned to different areas in the model.

Only minor changes in the output of the stage 1 models were evident. Two further AQUA models were defined, one simulating the wall as a thin low permeability area with nodes on both sides and one only using the area within the wall. Both indicated that boreholes set near the wall reduced water levels there by the required amount whilst levels in the centre of the site were largely unaffected.

The second stage also comprised inflow and upflow calculations into the contained area as a result of the difference in piezometric head between the underlying Sherwood sandstone (approximately 44m AOD) and the gravels. Further refining of the model and analysis with SEEP and AQUA showed that the total flow of groundwater to be abstracted is around 30l/s.

In order to limit the possibility of contamination migrating within groundwater over or under the wall, a groundwater abstraction and treatment system has been installed. The capacity of the system has been dictated by a design based on water levels inside the wall being at or below those outside, so in the event of leakage, clean water will flow in rather than contaminated water flowing out.

SEEP, a finite element seepage programme, was used to calculate inflow under the wall. Model runs using key-in depths into the Mercia mudstone of 0.5, 1, 2 and 3m and a head difference of 1m either side of the wall demonstrated that the flow under the wall decreased by about 20% with each increase in key-in depth. The 1m key-in depth selection stemmed from cost, buildability and inflow rate.

Groundwater abstraction system and treatment plant

A ground water abstraction system was installed to collect the water inside the wall along the river and railway line boundaries and currently consists of eighteen abstraction boreholes linked by a ring main to the groundwater treatment plant. The treatment plant was constructed on the southern side of the site adjacent to the River Derwent and railway line. The plant uses a series of clarification, biological and filtering processes to treat the contaminated groundwater, before it is discharged into the River Derwent. The effluent from the plant has to comply with the discharge consent conditions set by the Environmental Agency (EA). A system of continuous on-line monitoring is included within the plant to assist in complying with the EA requirements.

The abstraction system also includes nineteen pairs of monitoring wells located on either side of the containment wall. The wells provide access to monitor both the ground water levels and the degree of contamination present in the water. Both are required to prove the integrity of the bentonite wall.

Construction of the abstraction system and water treatment plant began in February 1997 by the Purac-Morrison Consortium (formed by Purac Limited, part of the Anglian Water group and Morrison Construction Limited) under a Design and Build Contract. On completion of the commissioning period (December 1997) the plant will be operated for a period of 15 years by Alpheus Environmental Limited (also part of the Anglian Water group).

Both the quantity of water to treat and the level of contamination present were assessed, but owing to the difficulty in obtaining information about the underlying ground water aquifer and the highly mobile nature of the contamination both have been difficult to predict. Thus the treatment plant was designed as a modular system to treat the assumed requirements of 33 l/s and for the level of contamination to be below maximum levels set out for all parameters included in the Discharge Consent. Space was provided however, to bolt on additional treatment systems and to increase the capacity of the plant, should more water need to be treated or the level of contamination in the water change with time. However in the long term the plant is expected to reduce the amount of water-soluble contamination present within the wall and it may be possible to remove some of the process systems.

Environmental monitoring

During the wall's construction, noise and dust were monitored, the latter at four places in the surrounding residential areas; this indicated that both remained within limits during the works. Gas monitoring also showed that elevations of gases were generally isolated and located within the landfill area. This monitoring is ongoing. River monitoring was undertaken at four points along the length of the whole site, over a 19-month period, including inorganic and organic analysis. No discernible trends were noted, either before or after wall construction.

Groundwater monitoring, comprising water level and chemical analysis, began in 1995 over the entire site in a limited number of boreholes. More regular monitoring commenced in May 1994 in boreholes inside and outside the wall, in particular where it was near the river. More recently water levels were monitored by the ground water treatment operators Alpheus and pumping now ensures that water levels inside the wall are maintained at or below those outside the wall. However at some points adjacent to the river this could be reversed due to hydraulic effects of the river.

Groundwater contamination, including organic and inorganic parameters, has been monitored in detail in boreholes inside and outside the wall since construction began in May 1994. A more comprehensive suite of tests including "Red List" substances, furans and dioxins was undertaken on fewer boreholes on a limited number of occasions. The main elevated parameters were boron, COD, ammoniacal nitrogen, sulphate, sulphide and electrical conductivity, although there was also significant areas of elevated polyaromatic hydrocarbons. Chemical monitoring continues around the wall in part of the groundwater treatment plant contract.

Conclusion

The Pride Park project has successfully implemented the Arup reclamation strategy including the construction of the bentonite cement containment wall, a waste repository, a ground water abstraction and treatment system. This allowed for approximately 75% of the Pride Park site to be developed, notably including the reclamation works for Derby County Football Club, buildings for various end uses such as a banking call centre, a hotel, restaurants, offices and a David Lloyd Leisure Centre. Further areas of the site are being currently reclaimed to greatly enhance this inner city area.

Construction on fill

H. Skinner
BRE, Garston, Watford, Herts.WD25 9XX

Introduction

The current emphasis on locating building developments on brownfield sites, many of which contain substantial depths of fill, means that the behaviour of fills which support the foundations of buildings or other construction is of increasing importance. Fill or made ground, consisting of material formed by a process of mechanical placement and/or compaction, poses particular problems for construction. This paper presents a brief outline of some of the difficulties encountered when building on fills, illustrated by case studies which demonstrate a number of features of fill behaviour.

Engineered fills should be distinguished from non-engineered fills. Engineered fill is selected, placed and compacted to an appropriate specification in a well controlled, monitored and documented process in order to achieve some required engineering performance. Non-engineered fills arise as the by-product of human activities associated with the disposal of waste materials and have not been placed with a subsequent engineering application in view; little control may have been exercised during placement and consequently there is the possibility of extreme heterogeneity and it is difficult to characterise the engineering properties and predict behaviour. In the past, many fills on which construction has taken place have been non-engineered and appropriate site investigation and ground treatment have been critical issues. Fill that has been engineered to an inappropriate specification, or has been placed in a poorly controlled compaction process, may have properties that are significantly poorer than anticipated, but should be less heterogeneous than a non-engineered fill.

Problems of building on filled ground can be separated into those arising as a result of:

- the state of the material – density, moisture content and history
- the geometry of the filled excavation or embankment

Most problems which arise when building on fills concern excessive settlement rather than low bearing capacity, although both are covered in this paper. In each section the fill behaviour and the situations in which this is likely to cause a problem are described and illustrated by laboratory results and field measurements.

A site investigation may be expected to define the moisture content, grading and in some cases the in situ density of the fill material. Compaction tests can then relate the density of the material to a standard value. The current state of the fill in terms of density and moisture content may be as a result of engineered or non-engineered placement. This paper deals with behaviour that might be expected based on the current state; improvement effected by ground treatment is not discussed.

Shear strength and bearing capacity

Granular fills

The permeability of even a well compacted granular fill will be sufficiently high that the drained angle of shearing resistance is the significant strength parameter. Well compacted, engineered granular fills will have a very high angle of shearing resistance that is composed of the constant volume value ϕ'_{cv} and an element that is dependent on dilation. The component that can be attributed to dilation is governed by the relative density (Bolton 1986) and may be as high as 10° but will reduce to zero at large strain or under very high stresses.

A non engineered granular fill is likely to be in a loose condition, with an angle of shearing resistance close to the constant volume value ϕ'_{cv}. This is usually similar to the angle of repose and is likely to be in the range 32° to 37° for sandfills and 35° to 42° for rockfills. Bearing capacity is therefore unlikely to be a problem for the typical load applied by a shallow foundation for low rise construction.

Clay fills

The undrained shear strength of a near saturated clayfill will govern behaviour when loading changes more rapidly than pore pressures can dissipate. The loading may be applied by a building or by self weight during placement. The undrained shear strength is primarily a function of moisture content and liquid limit. Figure 1(a) shows the variation of c_u with moisture content for a boulder clay compacted by 2.5kg and 4.5kg rammer tests; the vane shear strength varied between 20 and 150kPa (which was the upper limit of the apparatus). Only below 40kPa is the strength likely to cause significant problems associated with low bearing capacity. Strengths of this order are found in clayfills with a moisture content somewhat greater than w_{opt}.

The drained shear strength of a non-engineered clayfill is likely to be close to ϕ'_{cv} and in the range 20°-30°. Long term bearing capacity may be a problem only at the lower end of this range. Residual strengths of clay fills of high plasticity (I_P greater than 27%) may be very low, in the region of 10°. Figure 1(b) shows peak strengths derived from triaxial tests on compacted clays.

(a)

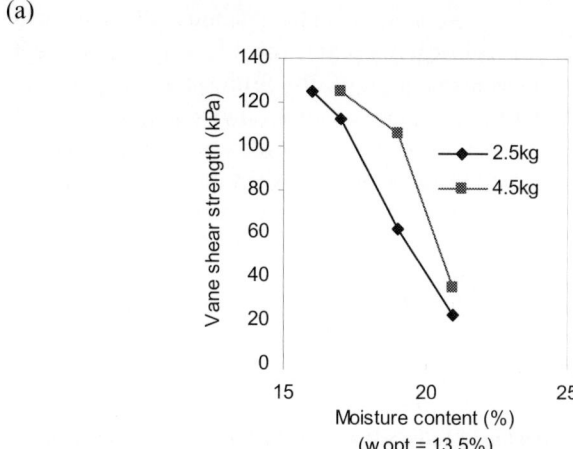

(b) after Vaughan *et al.*, (1978)

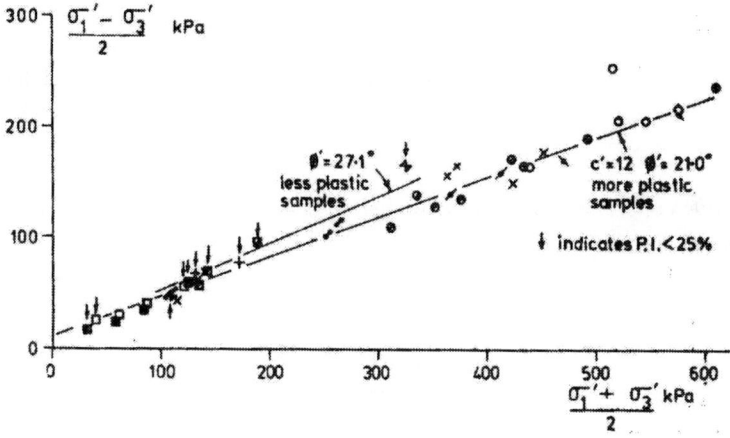

Figure 1 (a) undrained shear strength of compacted boulder clay fill and (b) drained peak strength of some compacted clay fills

Mixed fills

A fill that has significant proportions of both clay and granular material should be treated more as a clayfill if clay or silt size particles account for more than around 35% of the percentage by weight. Consideration should then be given both to the undrained shear strength and drained behaviour.

Settlement due to increased load

Self weight
With granular fills or any poorly compacted partially saturated fill, settlement due to the self weight of the fill will largely occur as the fill is placed, hence it will have little significance for construction on fills. With saturated or nearly saturated clay fills, dissipation of excess pore pressures may occur over a long period due to the low permeability of the clay and consolidation settlement may still be significant when construction takes place. Values for compressibility, that determines the primary settlement, are discussed in the next section.

Increase in applied stress
The short and medium term behaviour of fills in one-dimensional compression can be described by the constrained modulus, D, which is the ratio of applied vertical stress, σ_v, to the resulting vertical strain, ε_v. The stress-strain properties of fills are usually non-linear and the value of the constrained modulus will depend not only on initial density and moisture content but also on stress level and stress history. The compacted state of a clayey or granular fill can be described by 'compaction ratio', C_R, the ratio of dry density to the dry density achieved by compaction at optimum water content in the 2.5kg rammer test. At any particular stress level, the tangent constrained modulus, D_{tan}, will not be the same as the secant constrained modulus, D_{sec}. The secant constrained modulus can be related to any specified stress increment, but in this paper it is defined as the ratio of the applied vertical stress, σ_v, to the total vertical strain, ε_v.

The one-dimensional compression behaviour on first loading of a wide range of types of saturated fill and also dry sand fills is significantly non-linear and can often be approximated to a relationship of the form (Charles and Skinner 2001b):

$$\varepsilon_v = A_2\sigma_v^{0.5} \tag{1}$$

This relationship indicates that compression curves are parabolic in form which clearly cannot hold at very low stresses where such a relationship would imply infinite compressibility. For the relationship defined by equation (1), D_{tan} is twice as large as D_{sec}:

$$D_{sec} = \sigma_v/\varepsilon_v = \sigma_v^{0.5}/A_2 = D_{tan}/2 \tag{2}$$

The normalised constrained modulus (Schanz and Vermeer 1998), D_{secn} at a specified vertical stress $\sigma_{vn}=0.1\,\text{MPa}$ is then:

$$D_{sec} = 3.16\,D_{secn}\sigma_v^{0.5} \qquad \text{(N.B. all in MPa)} \tag{3}$$

D_{secn} is plotted against c_u for saturated fills in Figure 2(a). There is an approximately linear relationship between D_{secn} and log c_u for saturated clay fills and between compaction ratio, C_R, and log D_{secn} for saturated granular fills in Figure 2(b).

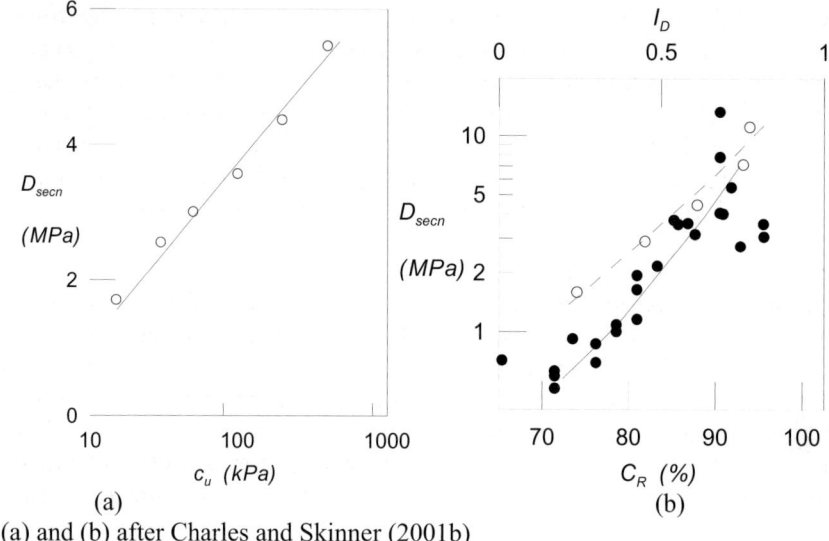

(a) and (b) after Charles and Skinner (2001b)

Figure 2 (a) D_{secn} vs c_u for a saturated clay fill and (b) D_{secn} vs C_R for two saturated granular fills

In the field most fills, particularly near the ground surface, will be in a partially saturated state. Oedometer tests with applied vertical stresses of up to 400kPa on compacted partially saturated samples of three fill types commonly found on brownfield sites have shown that the one-dimensional behaviour can take two different forms:

- With high air voids and at a moisture content dry of standard Proctor optimum, there is often a linear stress-strain relationship. Values of constrained modulus of 10 MPa or more have been measured on different fill types.

- With low air voids and wet of standard Proctor optimum moisture content, vertical strain typically is proportional to $\sigma_v^{0.5}$. For granular fills, the constrained modulus increases with increasing initial density index and for cohesive soils with increasing initial undrained shear strength. For both a coarse-grained colliery spoil and a boulder clay, D_{secn} is typically in the range 1-5 MPa.

Figures 3(a) and (b) illustrate the transition between the two types of stress strain relationship over this range of applied vertical stress, for a granular and a clayey opencast backfill.

(a)

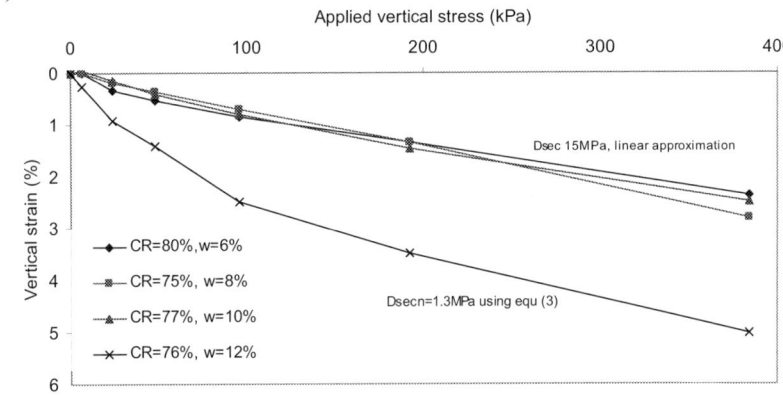

(b)

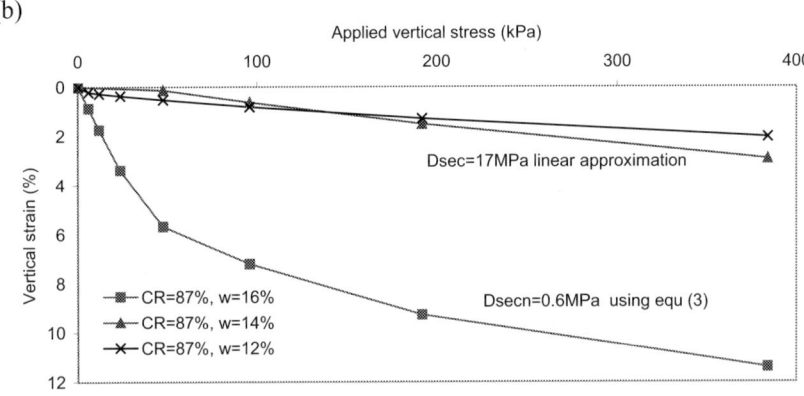

Figure 3 One-dimensional stress-strain behaviour for (a) granular colliery spoil fill (w_{opt}=7%) and (b) opencast mudstone backfill (w_{opt}=15%)

Long term 'creep' settlement

Self weight
Long term settlement due to self-weight which continues under conditions of constant applied stress and moisture content is more significant for construction taking place on fills than immediate self weight settlement. The linear relationship between 'creep' compression and the logarithm of the elapsed time is well established in the field as well as in laboratory tests. The field creep

compression rate parameter, α, which is analogous to the laboratory parameter C_α, is defined as follows (with α in percent):

$$(\alpha/100) = (s_2\text{-}s_1)/\{H \log(t_2/t_1)\} \tag{4}$$

where $(s_2\text{-}s_1)$ is the settlement of fill of height H, occurring between time t_1 and time t_2 since fill placement. A range of values for α determined by field measurements of the long term self weight settlement of fills is shown in Table 1. It should be noted that field measurements may not have been made under fully one-dimensional conditions.

Fill type	α (%)
Compacted sandstone/mudstone rockfill	$0.13\sigma_v{}'$
Uncompacted sandstone/mudstone rockfill	0.9-1.5
Uncompacted sandstone	1.2
Compacted sandstone	0.5
Compacted sandy gravel fill	$0.04\sigma_v{}'$
Compacted stiff clayfill	0.5
Puddled clay	1.0

N.B. ($\sigma_v{}'$ in MPa)

Table 1 Measured values of logarithmic creep rate (Charles 1993)

An uncompacted opencast backfill of 30m deep, with α=1%, might settle 90mm in the second year, in the 10th year the settlements would have reduced to 12mm/year and would continue to reduce. After 25 years the ongoing settlements should be negligible.

Increase in applied load

Long term settlements under the applied loads of a building may or may not cause problems.

In general, the total settlement under an increase in load can be described by a relationship of the form:

$$s = s_i + s_\alpha \log(t/t_i) \tag{5}$$

where s_i is the initial settlement over time t_i during and after loading. The time interval t_i over which the initial settlement is measured may be taken as of the order of 0.1 days in a load test where loading is rapid, and 10 days or more for the construction of a building. The parameter s_α is then analogous to α and is given by:

$$s_\alpha = (s\text{-} s_i)/\log(t/t_i) \tag{6}$$

Results from field load tests carried out at various sites are shown in Table 2, with the settlement divided by the width of the footing in order to non-dimensionalise it. Vertical stresses applied were of the order of 24-35kPa. The data can be used to predict the short and long term settlement under a footing load.

Fill type	S_i/b (%) t_i=0.1days	S_α/b (%)
Colliery spoil, placed by scrapers	1.2	0.3
Urban fill – sandy clay with brick fragments	0.7	0.1
Soft clay fill	0.2	0.1
Building waste	0.3	0.1
Alluvial sand (similar to sandfill)	0.2	0.04

Table 2 Measured values of logarithmic creep settlement (Charles and Skinner 2001b)

There are difficulties in measuring creep in oedometers when movements become very small; the least perturbation may cause much greater movements than those occurring due to creep. Side friction can increase with time and long term measurements may be suspect unless carried out in specially designed equipment. It can be concluded that it is difficult to obtain reliable predictions of creep settlement rates in the field from standard laboratory oedometer tests and predictions based on laboratory tests may substantially under-estimate field movements.

Settlement or heave caused by changes in moisture content

Collapse compression
Field investigations have demonstrated that collapse compression is a major cause of settlement of fills. The settlement can occur as the ground-water table rises or as surface water penetrates into the fill via shallow excavations through the surface crust.

- The effect of a rising ground-water table on the settlement of a mudstone and sandstone fill placed without any systematic compaction has been investigated at Horsley in Northumberland (Charles et al., 1977; Charles et al., 1993). Where the fill was 63 m deep, 0.33 m of settlement occurred and vertical compressions locally were as large as 2%.

- The effect of water penetrating into an uncompacted clay fill via surface trenches has been investigated at Corby in Northamptonshire (Charles et al., 1978; Burford and Charles 1991). A maximum surface settlement of

0.3 m has been recorded and vertical compression at one location has been as large as 6%.

For loose fills the movements due to collapse compression on inundation are likely to be much larger than those due to creep. Compaction ratio, C_R, and normalised water content $(w-w_{opt})/w_{opt}$, have been used to demonstrate the contours of collapse potential for a granular and clay fill (Figures 4(a) and (b)).

- Dry of standard Proctor optimum moisture content, collapse potential is largely a function of C_R. Typically $C_R = 95\%$ may be just adequate to eliminate collapse potential for a granular fill, but $C_R =105\%$ may be required for a clay fill.

- Wet of standard Proctor optimum moisture content, collapse potential is largely a function of air voids, V_a. Typically for a clay fill it will be necessary to reduce V_a to 5% to eliminate collapse potential, whereas for coarse-grained colliery spoil there may be very little collapse potential with V_a less than 10%.

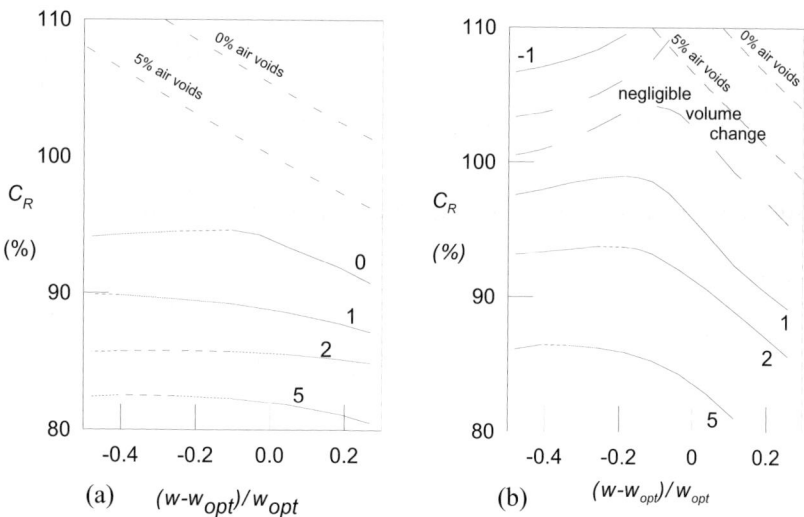

(a) Granular colliery spoil inundated at 60kPa
(b) Clay fill inundated at 30kPa
(a) and (b) after Charles and Skinner (2001b)

Figure 4 Contours of collapse compression on inundation in % vertical strain

Heave/shrinkage

Fill that is partially composed of clay with moderate or high volume change potential may show heave during increase in water content or shrinkage when dried. Heave on inundation with water can be seen in Figure 4(b) for the highly compacted samples of slightly expansive boulder clay fill. Lawton *et al.,* (1989) showed that the heave associated with wetting of a moderately expansive clay fill was related to the degree of compaction, water content and applied vertical stress. Greatest heave (of up to 10%) was shown by very dry, heavily compacted samples wetted at low applied vertical stress. Change in volume of the clay mineral is in part related to plasticity in the same way as natural shrinkable clays, although the actual volume change that can occur for a given change in water content will also be related to the compacted structure.

Fill behaviour – summary

The basic plot of dry density against moisture content and the results from the 2.5kg standard Proctor compaction test can be used to distinguish between different categories of fill behaviour that may be anticipated. In general, the further from $C_R=100$ and $w=w_{opt}$ the worse the problems of high compressibility, heave or collapse may be.

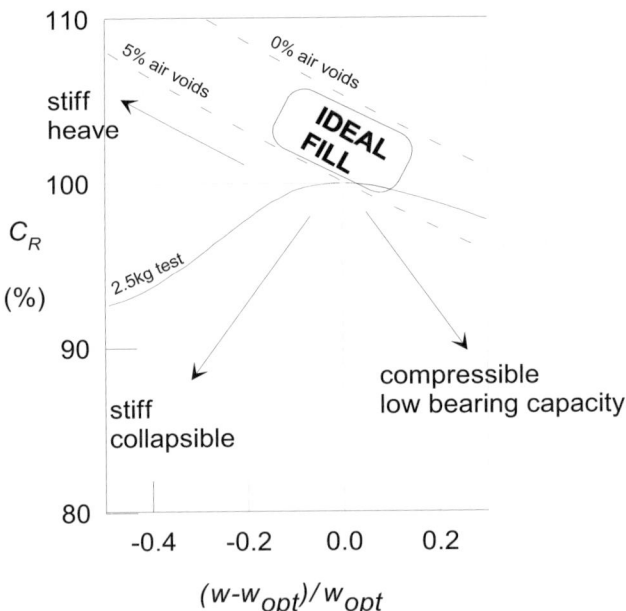

Figure 5 Summary of fill behaviour

Problems which arise as a result of the geometry of the filled void

Whilst a fill may be expected to perform adequately, based on measurements of its current state, the geometry of the filled void may result in differential settlements which are unacceptable. A framework for deciding where there may be a problem due to the fill geometry is outlined here.

The geometry of a cross-section of a backfilled excavation can be defined by the following parameters in which the term 'highwall' has been used to describe the slope irrespective of its steepness and height:

- height of highwall (H)
- depth of burial of top of highwall ($D \geq 0$)
- angle of highwall to horizontal (β)

In delineating the area of land which must not be built on, two primary factors control ground movements:

- the geometry of the highwall, which can be described by D, H and β,
- the volume reduction potential of the fill, which can be expressed in terms of a vertical compressive strain, ε_v.

Most field situations can be approximated to plane strain.

The criteria for tolerable deformations of buildings and infrastructure have a critical effect on the size of the exclusion zones. When building on fill, many acute problems arise with small buildings for which deep foundations are not an economically viable solution. For this type of building it is usually feasible to provide a stiff raft foundation which will resist distortion of the building due to differential settlement and horizontal tensile forces. The principal interest then lies in the delineation of the zone from which buildings should be excluded because tilt is unacceptably large.

A simple model of surface settlement, assuming that the tilt either varies linearly with horizontal distance or is constant, has been proposed by Charles and Skinner (2001a). The model enables the exclusion zone to be determined, based on the geometry of the void, acceptable tilt and likely fill compression.

The volume change in the fill that results in differential settlement is assumed to be associated with settlement of the fill below the level of the top of the highwall and can be expressed as $\varepsilon_v = s_M/H$, where s_M is the maximum settlement. The resulting surface settlement profile consists of linear and parabolic sections. The width of the area over which tilt is non-zero can be expressed in terms of a limit angle, γ, and the wall geometry. The limit angle defines the extent of the differential settlement away from either the top or base of the wall. Where the slope is long, there is an area defined by the limit angle over which the tilt is constant. Where the slope is short, there is no such area and the point of inflection in the surface settlement profile can be related to the wall geometry. The maximum tilt can be found by integrating the tilt profile such that it equals the maximum settlement. The proposed model could also be used to predict other parameters for ground distortion, but these are not discussed here. For example, in the case of a short buried slope, the model prediction for surface settlement and tilt is shown in Figure 6.

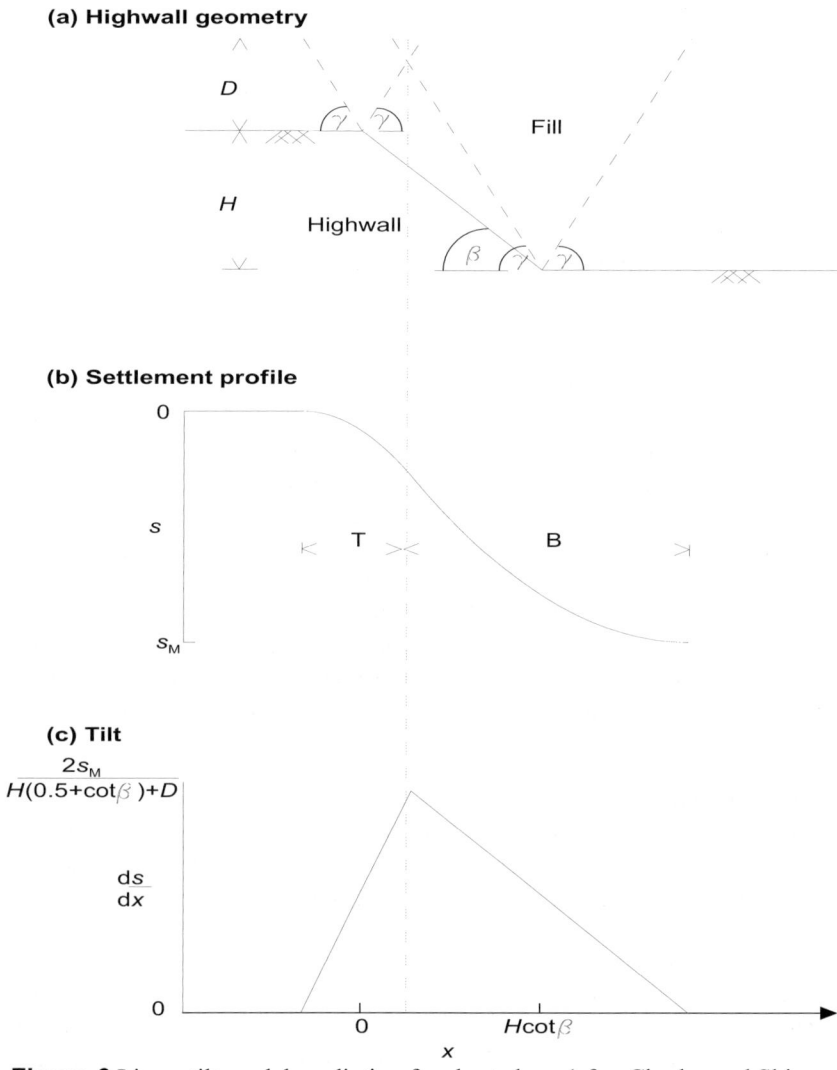

Figure 6 Linear tilt model prediction for short slope (after Charles and Skinner, 2001a)

A small value of γ implies a large area of relatively small tilt whereas a large value of γ implies larger tilt affecting a smaller area. The value of γ that has been adopted (63.4°, corresponding to $\cot\gamma=2$) is towards the upper bound of probable values and therefore should give a moderately conservative value for maximum tilt. However, it could slightly under-estimate the area affected by tilt.

In order to predict the position and width of exclusion zones; the engineer must decide on an acceptable value of tilt, α_T and the potential vertical compression in the fill below the top of the highwall ε_v. Charts have been produced that relate maximum tilt α_M for given values of ε_v and wall geometries, and for sites where tilt will be a problem the location and size of the exclusion zone for these parameters can be identified Charles and Skinner (2001b).

Case studies
Two case studies will be used to illustrate some of the aspect of fill behaviour that have been described in the earlier sections.

Case study 1: Fill behaviour: Blindwells Opencast Site.
Where a trunk road was to be built across the Blindwells opencast site near Edinburgh, an embankment was constructed in which the top 16m of the 60m deep mudstone, siltstone and sandstone backfill were systematically compacted. The rest of the void was filled with uncompacted fill. Settlements with depth have been measured since 1984, after fill placement, both where the upper fill has been compacted and in an area of uncompacted fill (Charles 1993). Surface settlements have been monitored over an 8 year period across the highwall of the excavation, along the compacted embankment.

Fifteen years of monitoring at Blindwells Opencast Site have given data on the basic material behaviour including the difference in long term settlement between fill placed in engineered and non-engineered operations and movements due to a recent sudden change in water level.

Creep settlement
In 10 years the area with compacted fill settled 0.2m with virtually no compression in the upper compacted zone. The uncompacted fill away from the embankment settled 0.44m with an α value of about 1%.

Collapse compression
The water level in the fill has been maintained at a low level by pumping until recently. In that time no significant changes in water level had taken place. A recent increase in water level resulted from turning off the pump in the area and allowing the water table to return to the pre-excavation level. Movements due to the rise in water level have been monitored in the uncompacted fill. The site

topography was such that no water level rise occurred at the location of the magnet extensometer in the compacted fill. Surface settlement of 250mm occurred during the period when the water level rose by 13m through the uncompacted backfill at the location of the magnet extensometer. Figure 7 shows that the compression occurred within the backfill where the water level rose and just above it. The compression in the newly wetted fill corresponded to 1.4% in fill that had only compressed by 0.2% over the previous 18 months.

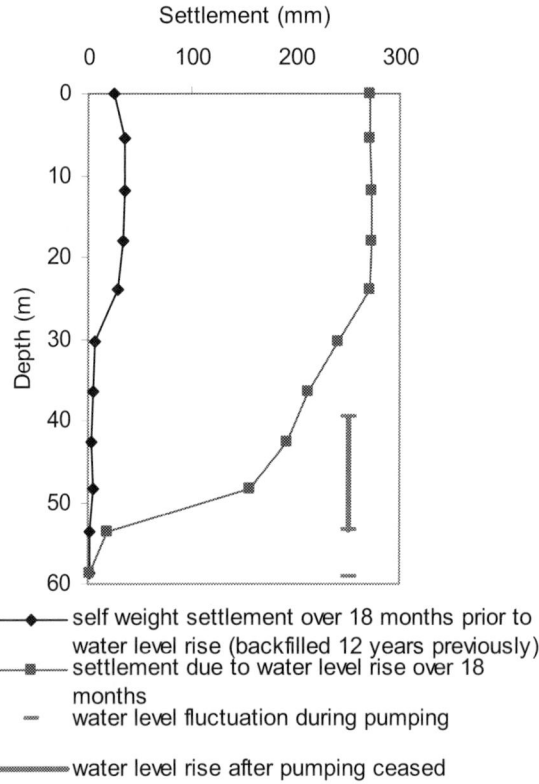

Figure 7 Settlements measured in uncompacted backfill at Blindwells

Case study 2: Changes in fill geometry: Orgreave Opencast Site

Settlement has been monitored over a highwall ($\beta = 38°$, H=50m and $D = 0$) at this opencast mining site near Sheffield (Charles and Skinner 2001a). The creep settlement of the end-tipped mudstone rockfill which has occurred between May 1999 to June 2000 is shown in Figure 8.

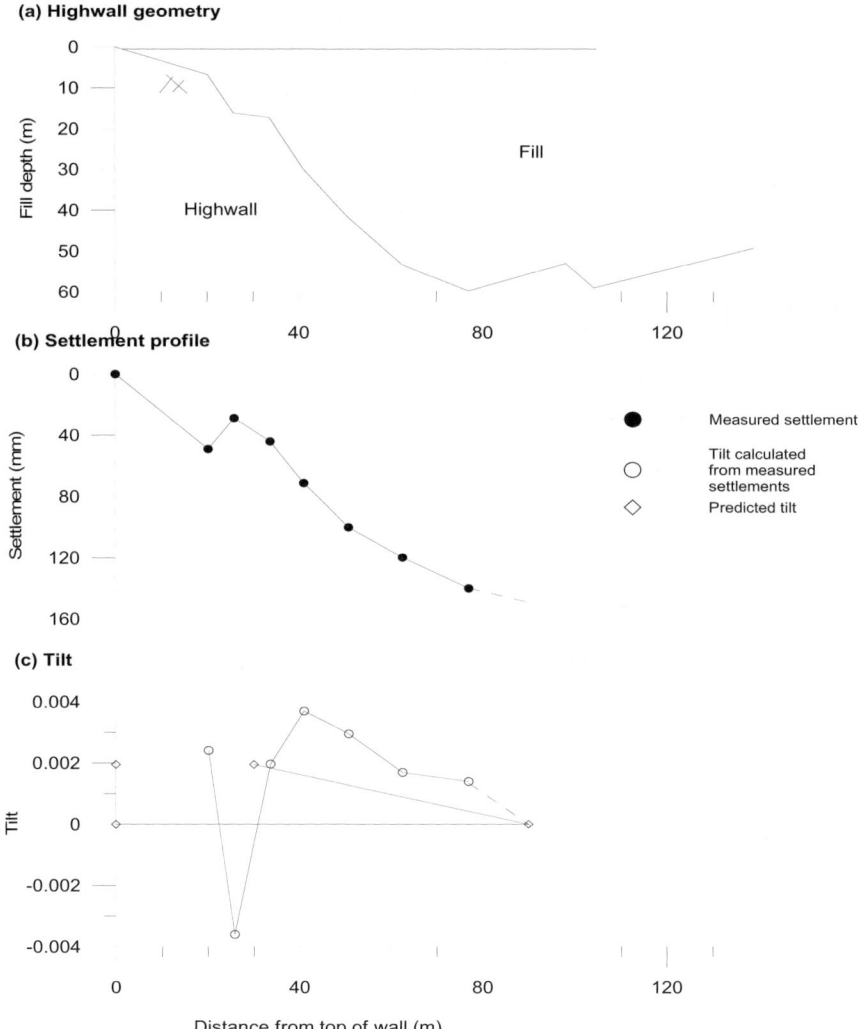

Figure 8 Settlements measured at Orgreave opencast site (after Charles and Skinner (2001b)

It would seem that a ledge close to the top of the highwall has caused a local reversal in the tilt. Nevertheless, away from this anomaly there is reasonable agreement between the tilt predicted by the linear tilt model and that derived from the measured settlement. The presence of an overburden heap prevented settlement observations being made at a distance from the toe of the highwall. Significant settlements, and differential settlements have occurred in the vicinity of the highwall, over a relatively short monitoring period whilst the site is still operational. Continuing settlements due to self weight will reduce over time, and in most of the site by the application of a significant preloading due to the overburden heap. The preloading may also reduce any collapse potential in the fill.

Conclusions

This paper has highlighted some problems that may arise when building on fills, as a result of the material state or fill geometry. Generally, unless fill is placed very loose or wet, bearing capacity is unlikely to be a problem and settlement will form the main criterion for acceptable performance:

- When loaded under one dimensional conditions, the primary compression of a saturated or near-saturated fill under drained conditions can be approximated by a power law in stress-strain in which the strain is proportional to the square root of the applied stress. Fills with higher air voids are likely to show a more nearly linear stress-strain relationship with higher constrained moduli.

- Settlement can also occur without changes in applied stress. Values for creep rates that can be expected for different types of fill have been given. Fills with air voids greater than 10-15% may show large collapse settlements when wetted.

- A model for the identification of situations in which changes in fill placement geometry may result in excessive differential settlement has been outlined.

Whilst construction on fills is likely to be more difficult than construction on natural ground, provided that designers understand the controlling behaviour of the fill materials as placed they can design for the possible hazards.

Acknowledgements

The research work described in this paper has been carried out for DETR under the Building Regulations Framework Agreement and Sustainable Construction Business Plan.

References

Bolton, M.D. 1986. *The strength and dilatancy of sands*. Geotechnique, 36, (1), 65-78.

Burford, D. & Charles, J.A. 1991. *Long term performance of houses built on opencast ironstone mining backfill at Corby, 1975-1990*. Ground Movements and Structures. Proceedings of 4th international conference, Cardiff, 54-67. Pentech Press, London.

Charles, J.A. 1993. *Building on fill: geotechnical aspects*. BRE Report 230, CRC, London.

Charles, J.A., Burford, D. & Hughes, D.B. 1993. *Settlement of opencast coal mining backfill at Horsley 1973-1992*. Engineered Fills (eds B G Clarke, C J F P Jones and A I B Moffat). Proceedings of Conference, Newcastle upon Tyne, 429-440.

Charles, J.A., Earle, E.W. & Burford, D. 1978. *Treatment and subsequent performance of cohesive fill left by opencast ironstone mining at Snatchill experimental housing site, Corby*. Clay Fills. Proceedings of conference held at Institution of Civil Engineers, November 1978, 63-72. Institution of Civil Engineers, London, 1979.

Charles, J.A., Naismith, W.A. & Burford, D. 1977. *Settlement of backfill at Horsley restored opencast coal mining site*. Proceedings of Conference on Large Ground Movements and Structures, Cardiff, 229-251. Pentech Press, London.

Charles, J.A. & Skinner, H.D. 2001a. *The delineation of building exclusion zones over highwalls*. Ground Engineering, 34, (2), 28-33.

Charles, J.A. & Skinner, H.D. 2001b. *Compressibility of foundation fills*. Geotechnical Engineering. *(in press)*

Lawton, E.C., Fragaszy, R.J. & Hardcastle, J.H. 1989. *Collapse of compacted clayey sand*. ASCE Journal of Geotechnical Engineering. 115, (9), 1252-1267.

Schanz, T. & Vermeer, P.A. 1998. *On the stiffness of sands*. Pre-failure deformation behaviour of geomaterials (eds R J Jardine, M C R Davies, D W Hight, A K C Smith and S E Stallebrass). Geotechnique Symposium in Print, 383-387.

Vaughan, P.R., Hight, D.W., Sodha, V.G. & Walbancke, H.J. 1978. *Factors controlling the stability of clay fills in Britain*. Clay Fills. Proceedings of conference held at Institution of Civil Engineers, November 1978, 205-217.

Problematic soils or is it problematic specifications?

R. Coombs[1], L.K.A. Sear[2] and A. Weatherley[3]
1) *Innogy PLC, Selby, North Yorkshire*
2) *UKQAA, Wolverhampton*
3) *PowerGen, Ratcliffe, Nottinghamshire*

Introduction

Specifications generally develop from experience with well known materials and practices. Since many of the materials used are high quality, primary materials, the performance of which is well understood, the specifications reflect their performance. Whilst this has benefits in that there are standards against which these materials can be judged, the specifications will tend to result in the rejection of materials that might otherwise be suitable.

There are many advantages in using Pulverised Fuel Ash (PFA) in construction, in many instances it can have benefits over natural materials. However, there is a reluctance in some quarters to its use because it is either too novel (it has only been around for half a century), that it is a waste material and therefore must be lower grade, or simply that it does not fit the specification and is therefore unsuitable.

It is clear from many reports in its early days that PFA was a material suitable for many applications including engineering fill, and many millions of tonnes have been used over the last 50 years. However, despite all the successes there have been some problems with using PFA as a fill material, not because it was unsuitable but because it did not meet the relevant specification. It has often been the case that PFA has had to meet an unsuitable specification rather than a suitable specification being used to gain the benefits of PFA. This paper details the sort of problems that have been encountered in the past.

PFA as a fill

Early use of PFA

An early example (Sutherland, 1966) involved the filling of a six metre deep disused railway cutting on the A452 road at Packington. The PFA was dug from lagoons at Hams Hall power station and placed by end tipping. The moisture content of the PFA was reported as 55%. Despite this the settlement was only 38mm, all of which occurred during the first two months. A further example is the factory in Manchester, reported by (Raymond, 1966). The factory was built on 15 metres of stockpile PFA at Agecroft, the PFA had received no compaction other than that from normal spreading operations. Despite this the settlement measured when the factory was built was less than 20 mm.

Despite the success in the use of PFA as fill there was considered to be a need to control the density better. Further work (Margason and Cross, 1966) indicated that there could be significant variation in maximum dry density (MDD) with PFA. They commented "It therefore seems unlikely that a meaningful specification in terms of required dry density could be used even with ash coming from one known source, unless research showed that normal standards may be relaxed on this material". Thus it was accepted very early on that the criteria that were used for other fill materials were inappropriate for PFA.

However, despite the advice of Transport and Road Research Laboratory (TRRL), the Specification for Highway Works (Department of Transport, 1991) at present requires compaction on site to achieve a dry density at least 95 % of MDD. The introduction of this requirement has led to difficulties in controlling the compaction of PFA, as was concluded would happen by Margason and Cross. Their report showed quite a spread in both the maximum dry density and optimum moisture content of PFA. The issue is how significant this variation is in practice, remembering that the use of PFA had not led to significant control problems prior to this.

Factors affecting the density of PFA

There are a number of issues that affect the dry density of PFA including chemistry, grading, carbon content and "floater" content (that proportion of the PFA that can float on water). "Floaters", or cenospheres, are hollow glassy spheres with the same chemical composition as the PFA.

The chemistry is dependent on the coal burnt. The major elements in typical UK PFA are silicon, aluminium and iron, with lesser amounts of calcium, magnesium and alkali metals. Also an increase in the proportion of iron causes an increase in the density of the PFA.

The grading is affected by the grinding of the coal and the collection system in the gas stream. The greater the spread of sizes, the better the packing of the particles.

The carbon content, measured as loss on ignition (LOI), is a function of the milling of the coal and the air content at combustion. If a power station is base loading (that is, running 24 hours a day) the conditions are ideal and the carbon content will be low. If the running is more intermittent then there will be more carbon present. Increased levels of carbon results in an increase in the Optimum Moisture Content (OMC) and a reduction in MDD.

The floater content is less easy to quantify, but clearly the more floaters there are the lower the density will be, although the effect will be small.

Air voids assessment

One method of assessing the degree of compaction of a soil is to measure its air voids content. This is not usually applied to PFA because the air voids content at maximum compaction tend to be higher than for typical soils. The carbon content of the PFA can affect the measured air voids content in PFA. The carbon particles are open-textured and will absorb a large quantity of liquid. Thus when they are tested for particle density the voids in the carbon will be filled and the density will be measured as being close to that of graphite, typically 2.51 Mg/m^3. However, when the dry density is determined these voids are empty and the carbon has an effective similar to that of coke (typically 1.35 Mg/m^3). If the density of the bulk of the PFA particles is assumed to average 2.3 Mg/m^3 then the effect of the different ways of expressing the density of the carbon can be seen in Figure 1.

The plot shows that if the density of the carbon is taken to be the same as graphite then the air voids line show a higher dry density for a given moisture content than if the density of the carbon is taken as that of coke. For a 10 % loss on ignition, a dry density of 1.35 Mg/m^3 and a moisture content of 25 %, the air voids are 8% and 5 % respectively. The effect is more marked for higher loss on ignition PFA. This demonstrates why air void content is not a suitable means of assessing compaction of PFA and have not been used as a control method by the Department of Transport (DoT).

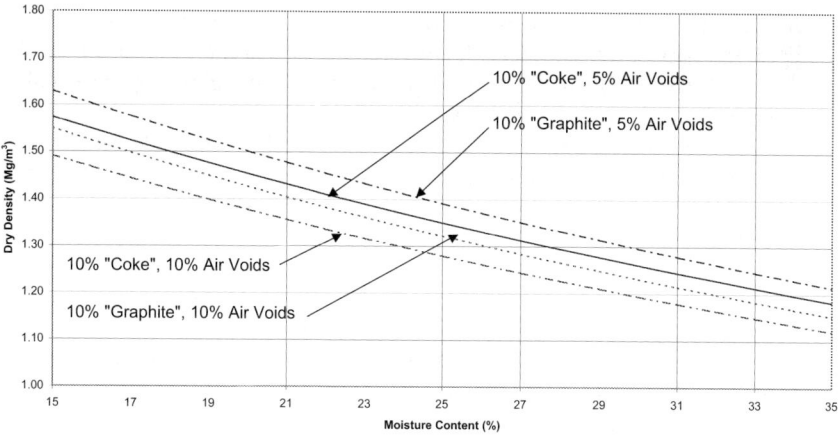

Figure 1 Variation in air voids with carbon type

Compaction compared to maximum dry density

The DoT specification treats PFA differently from other materials when it is used as a general fill requiring the compacted density to be at least 95 % of maximum dry density. Other general fill materials are controlled by a method specification. As was shown earlier, if the acceptance criteria for compaction of PFA are set too tightly then problems of density control will arise. An example of variability in maximum dry density is shown in Table 1; the PFA in this instance is one of the most consistent in the UK and the results are taken over a number of years. It shows that even this PFA is likely to present problems if a fixed target density is chosen. If the target is set at the mean MDD then there will be occasions when the best that is achievable is 97 %, and therefore the likelihood of failure is increased. The risk of non-compliance is even greater if the highest values were to be set as targets.

Mean MDD (Mg/m^3)	1.46
Range of MDD (Mg/m^3)	1.42-1.48
Range of MDD relative to the mean (%)	97-101
Range of MDD relative to the highest value (%)	96-100

Table 1 Variability of MDD for PFA

Although the specification would appear to show that the variation in the properties of the PFA is high this is not necessarily the case. This particular PFA, even without any processing, meets the requirements of BS EN 450 (BSI, 1995), the European standard for fly ash (PFA) for use in concrete. The typical data for this PFA compared to the relevant requirements of BS EN 450 are shown in Table 2. The PFA easily meets the stringent requirements for use in a high value product (concrete) but is considered as too variable for a low value use (general fill).

The data suggest that the most consistent PFA are as consistent as many manufactured products but that its variability is too high to meet the 95 % of MDD target. Less consistent PFA's will have an even greater problem, although past experience has indicated that this is not a problem in reality. The conclusion must be that the target set is not appropriate, rather than that the material being too variable.

Parameter	Typical range	Requirement of BS EN 450
Loss on ignition (%)	2.5-4.5	<5.0
Fineness (% retained 45μm sieve)	23.5 ± 5	± 10 from mean
Particle density (kg/m^3)	2160 ±100	±150 from mean
28 day activity index (%)	77 – 87	>75
90 day activity index	85 -93	>85

Table 2 Typical data to BS EN 450

Environmental issues

Another area where problems have occurred is with environmental concerns, especially trace metals. Again the problems have arisen through a lack of understanding of PFA.

Most soils are produced by erosion of rocks followed by deposition by ice, water or air. These can subsequently be contaminated particularly through the actions of man. PFA is produced in a furnace where the minerals become molten in the flame and the droplets are rapidly cooled as they pass out of the furnace. This results in particles that are mainly amorphous glass in which is entrained most of the elemental constituents that were present in the coal. A small proportion of the elements is deposited on the surface and these are available for leaching.

Another issue to be considered is the permeability of the PFA. This is typically 10^{-7}m/s, which is intermediate between clay and sand, but low enough to restrict the rate of flow of water through the compacted material. BRE Digest 363 (BRE, 1991) considers soils with permeability of less than 10^{-5}ms^{-1} to have a low enough permeability to reduce the sulfate exposure class for concrete by one class.

One element that has been perceived as a particular problem is arsenic. The Inter-departmental Committee on Reclamation of Contaminated Land (ICRCL) guideline threshold for arsenic in land to be used for parks, playing fields and open spaces is 40 mg/kg. The amount of arsenic in PFA is usually in excess of 50 mg/kg. However, as discussed above, most of the trace elements are entrained in the glass and are not available. If the leachable fraction is considered, rather than the total, then the situation changes. The accepted leaching test involves extraction using a 10:1 water/solids ratio. If this method of extraction is used then the leachable arsenic is measured at less than 1 mg/kg, below the threshold for soils for use in domestic gardens and allotments. In other words there is no risk from arsenic in PFA.

Conclusions

When considering the use of a *novel* material for any application it is important to understand its properties and build a specification around its actual performance. Imposing unsuitable specifications on materials will ultimately lead to a reduced use of alternative and recycled materials.

References

BRE. 1991. *Sulphate and acid resistance of concrete in the ground.* Digest 363. Watford.

BS EN 450: 1995. *Fly ash for concrete – Definitions, requirements and quality control.* British Standards Institution, HMSO, London.

Department of Transport. 1991. *Specification for Highway Works*, HMSO, London.

Margason, G. & Cross, J.E. 1966. *Settlement behind bridge abutments: Use of PFA on the Staines By-pass bridges*. Road Research Laboratory Report 48. Crowthorne.

Raymond, S. 1966. *Shear strength, settlement and compaction characteristics of pulverised fuel ash*. Civil Engineering and Public Works Review.

Sutherland, H.B., Finlay, T.W. & Cram, I.A. 1996. *Engineering and related properties of pulverised fuel ash*. The Journal of the Institution of Highway Engineers.

Ground improvement by dewatering

M. Preene
Arup Geotechnics, Leeds, LS9 8EE

Introduction

Dewatering is an established construction expedient whereby groundwater is controlled by some form of pumping. The most obvious aim of dewatering is to keep the working area 'dry' and free of water. Another benefit, sometimes overlooked, of lowering groundwater level is the increase in effective stress, which can dramatically improve the properties of problematic soils.

This paper outlines the improvement in soil properties that may result from dewatering or the reduction in pore water pressures. Some of the less commonly used methods are reviewed, including: vacuum ejector wells; vacuum-accelerated consolidation; and electro-osmosis systems. Some of the more unusual applications of dewatering are described.

Groundwater and instability

When excavating below the groundwater level in a permeable water-bearing soil, it is obvious that water will flow into the excavation, potentially resulting in flooding of the working area. Dewatering (which is a form of 'groundwater control') typically involves abstracting water from an array of wells or sumps in or around an excavation. This results in a local, temporary, lowering of groundwater levels around an excavation.

However, in addition to problems with flooding, excavations below the groundwater level often experience instability. This can include large-scale slope failures, smaller slumping or erosion features in slopes, or the heave, softening or loosening of the soil beneath excavation formation level. This groundwater-induced instability can be explained in terms of the effective stress regime around an excavation.

The principle of effective stress is expressed by Terzaghi's deceptively simple equation:

$$\sigma' = \sigma - u \tag{1}$$

Where σ is the total stress, u is the pore water pressure and σ' is the effective stress. The effective stress in the soil controls the ability of the soil to resist shear loading. This is illustrated by the Mohr-Coulomb failure criterion which shows that for a soil of angle of shearing resistance ϕ', the shear strength at failure τ_f is proportional to the effective stress:

$$\tau_f = \sigma' \tan \varphi' \tag{2}$$

These equations show that the abstraction of water and the lowering of groundwater levels reduces pore water pressures and will increase the effective stress in the area affected. This produces a corresponding increase in the shear strength of the soil, improving the stability of the base and sides of the excavation.

Dewatering can improve stability in a wide range of soils, but effects can be particularly dramatic in soils of moderate to low permeability, such as fine sands, silts and laminated clay/silt/sand mixtures.

These soils do not drain easily, and any excavation below the groundwater level will encounter only minor seepages, and is unlikely to flood rapidly. Yet, even the small seepages encountered (perhaps less than 1 l/s for a large excavation) can have a dramatic destabilising effect on the sides and base of an excavation. On site, people are often surprised that such small flow rates can be a problem, but effective stress theory explains the mechanism of instability. The seepages imply the presence of high positive pore water pressures around and beneath the excavation. This implies low levels of effective stress, and hence low soil strength – instability is the natural result.

A system of pumped wells will lower the pore water pressures around and beneath the excavation to maintain effective stresses at acceptable levels and prevent instability. The aim is not to totally drain the pore water from the soils – this would be very difficult as capillary forces mean that fine-grained soils can remain saturated even at negative pore water pressures. Because the soil is not being literally 'dewatered', pumped well systems in fine-grained soils are more correctly referred to as pore water pressure control systems; in these cases 'dewatering' is a misnomer.

The instability resulting from uncontrolled groundwater, and the improvements that can be achieved by reducing pore water pressures can be illustrated by considering two common cases – 'running sand' and 'quick' conditions.

Running sand conditions

'Running sand' is a colloquial term used to describe conditions when a granular soil (which can include silts and very sandy gravels as well as sand) becomes so weak that any slope or cut face cannot stand, and becomes an almost liquid slurry. The term is often used as if it were a property of the soil itself. Actually, it is the seepage of groundwater through the soil, and the resultant low values of effective stress, that cause this condition. Reduction of pore water pressures can change 'running sand' into a stable and workable material.

Figure 1 shows an excavation with sloping sides dug in a bed of silty fine sand below the original groundwater level. If the inflow water is pumped from a sump within the excavation, the sides will slump in when a depth of about 0.5 to 1.0 m below original water level is reached. As digging proceeds the situation will get progressively worse, and the edges of the excavation will recede. The bottom will soon fill with a sand slurry in an almost liquid condition which will be constantly renewed by material from the side slopes. The collapse of the side slopes results from the presence of positive pore water and seepage pressures resulting from the flow of water to the sump.

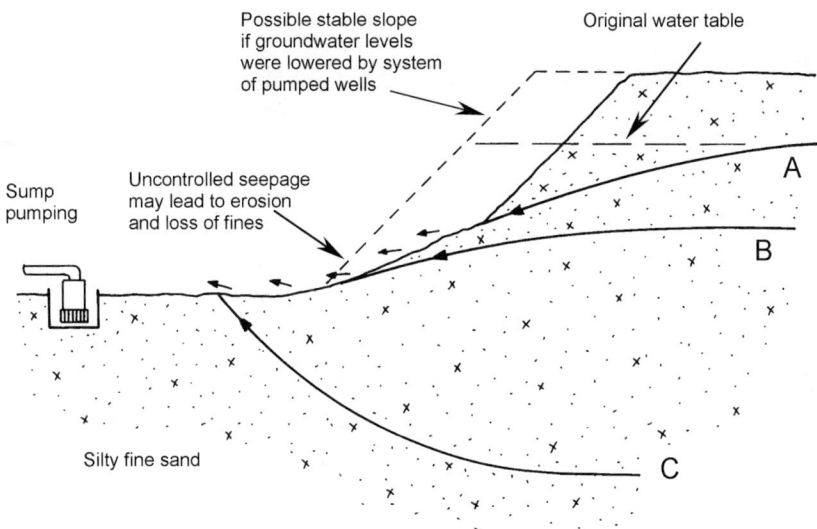

Figure 1 Running sand conditions

This instability can be explained in terms of effective stress. In Figure 1 flow line 'A' represents the flow line formed by the water table or phreatic surface. Seepage into the excavation will cause a slight lowering of the water table so flow line 'A' curves downward and emerges almost parallel to the surface of the excavation. Where the water emerges from the slope there will be positive pore water pressures, which reduce effective stresses in the soil. Immediately below the emergence of the flow line the pore water pressures generated by the seepage mean the soil can no longer support a slope of ϕ' which would be possible in a dry soil. Below the emergence of the seepage line the soil slope will slump to form shallower angles. At the point of emergence of the almost horizontal flow line 'B' the sand will stand at $\frac{1}{2}\phi'$ or less. Where there is upward seepage into the excavation base (flow line 'C') the effective stress may approach zero and the soil cannot sustain any slope at all. In these circumstances the so-called 'quick' case described below may develop.

Running sand conditions can be avoided by the simple expedient of installing an array of pumped wells around the perimeter of the excavation (Figure 2). The wells lower the groundwater level and prevent seepages emerging in the sides and base of the excavation. Thereby, positive pore water pressures are not generated in those areas, effective stress levels are controlled, and the soil does not lose its strength.

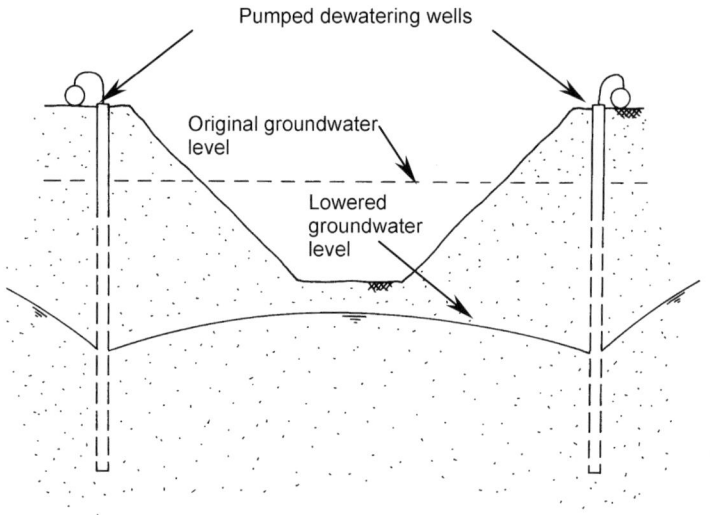

Figure 2 Pumped well pore water pressure control system

Quick conditions

If the sides of the excavation are supported by physical cut-off walls then it is possible to pump from within the excavation, without the risk of instability of side slopes. The risk of 'quick' conditions occurring in the base due to upward seepage remains (Figure 3). Quick conditions occur when high pore water pressures associated with upward seepage of water reduce the effective stress to zero. The soil will 'boil' or 'fluidise' and lose its ability to support anything placed on it.

Fluidisation will theoretically occur when the upward hydraulic gradient exceeds a critical value i_{crit}:

$$i_{crit} = (\gamma_s - \gamma_w)/ \gamma_w \tag{3}$$

where γ_s is the unit weight of soil, and γ_w is the unit weight of water. In general the density of soil is about twice that of water (peat being a notable exception), so typically $i_{crit} \approx 1$. This is the hydraulic gradient at which fluidisation will occur, so it is important that this value is not approached. In design it is normal to limit the predicted hydraulic gradients to be less than i_{crit}/F, where F is a factor of safety greater than unity.

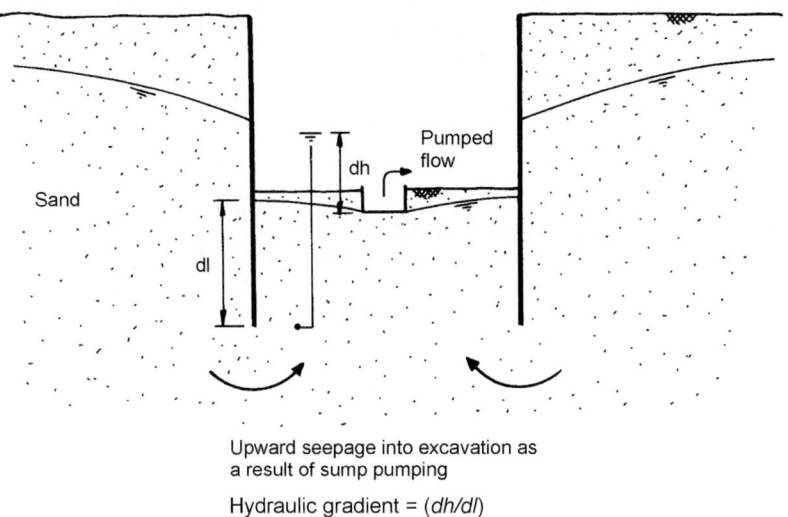

Upward seepage into excavation as
a result of sump pumping

Hydraulic gradient = (dh/dl)

Figure 3 Quick conditions

Excessive vertical hydraulic gradients can be avoided by various means. The flow path for seepage can be increased (by deepening the cut-off walls) or by reducing the head difference between the outside and inside of the excavation by installing a system of wells outside the excavation. Alternatively, the flow regime within the cut-off can be manipulated by the installation of wells to prevent vertical seepage occurring through the vulnerable soils immediately below formation level (Figure 4).

The dramatic improvements in stability that can be attained by reduction of pore water pressure can be illustrated by a case history. A small-diameter shaft was sunk by underpinning through a stratum of fine sand, with the soil being excavated by men working in the base of the shaft. When the water table was encountered, sump pumping was used to remove the relatively small volumes of water entering the shaft. By the time the shaft had penetrated a metre or so below original groundwater level the sand was literally 'running' into the bottom of the shaft. Despite several skips of spoil being removed, the excavation could not be taken any deeper – soil flowed in to replace that just removed. Sensibly, at this stage the shaft was flooded while a system of ejectorwells was installed to lower pore water pressures. This increased the effective stresses and stabilised the sand. The sand which was previously almost a slurry was now so strong it had to be excavated using air tools!

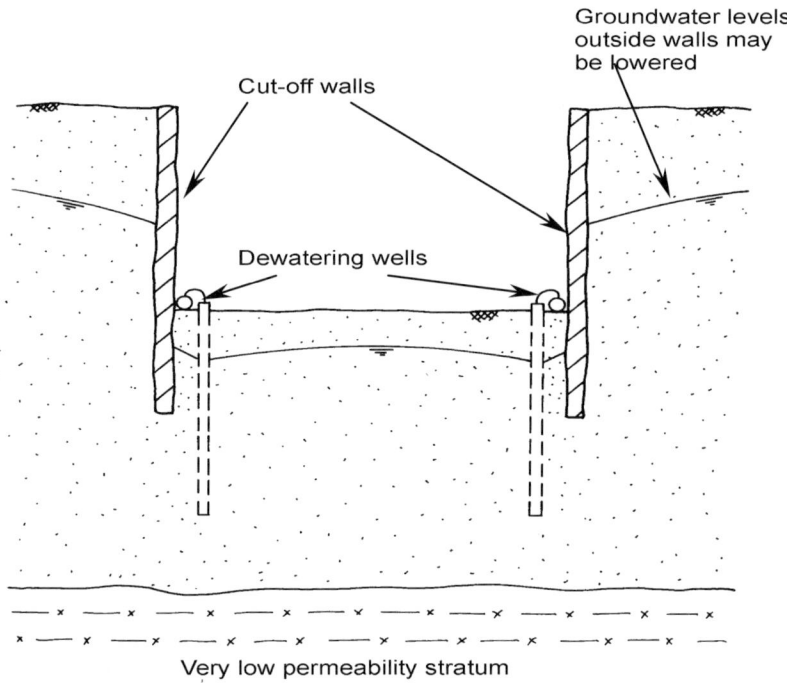

Figure 4 Use of pumped wells to control vertical seepage

Dewatering as a means of ground improvement

The effective stress changes that result from reductions in pore water pressures mean that dewatering is often a *de-facto* form of ground improvement. This is despite nothing being added to or mixed with the soil, and despite the soil not being physically re-arranged in any great way. This contrasts sharply with many other ground improvement methods that improve soil properties by injection (of grout or other material) or rearrange the soil (for example by densification).

When comparing ground improvement methods, it is interesting to note that, at present, the relative environmental impact of competing methods is rarely considered. It might be argued that the impact of the truly temporary methods such as dewatering and artificial ground freezing is relatively benign, since no grouts are additives are left in the ground on completion. The impact of dewatering and artificial ground freezing can be further differentiated by the energy consumption of each, amongst other factors: dewatering systems use little energy in comparison to freeze systems. The point of the comparison here is to illustrate that in the future we may need to assess the environmental implications of ground improvement methods. When we do, it may be possible to classify dewatering methods as being of low impact.

Specialist methods of groundwater control
A wide range of dewatering techniques are available (see Preene *et al.*, 2000) for a variety of applications. Some of the specialist techniques suited to true 'ground improvement' applications are described below.

Vacuum ejector wells
The vacuum ejector well method (also known as the eductor or jet-eductor system) uses an array of special pumped wells (Figure 5) to reduce pore water pressures in low permeability soils such as very silty sands, silts or clays with permeable fabric.

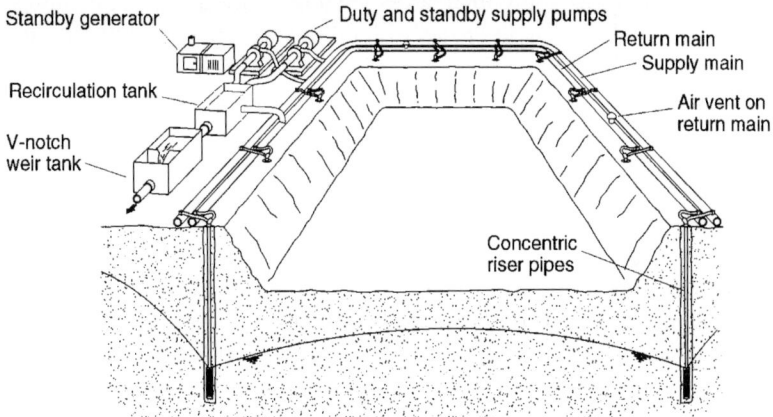

Figure 5 Ejector system components (from Preene *et al.*, 2000: reproduced by kind permission of CIRIA)

Each ejector well is pumped by an hydraulic device – the ejector – located within the well. The ejector comprises a nozzle and venturi which, when high pressure water is circulated through it, can pump both air and water. If the ejector capacity exceeds the rate that water is flowing into the well (as is often the case in low permeability soils), and if the top of the well casing is sealed, a vacuum will be generated in the well. This vacuum enhances drainage of water and reduces the pore water pressures in the soil.

The performance of vacuum ejector systems is described in detail by Powrie and Preene (1994). The technique is highly specialised and needs careful application. Particular points to note are the high energy consumption of ejector well systems in comparison with other pumped well methods, and potential problems with long-term operation of ejector systems. The nozzle and venturi in the ejector can be prone to clogging by bacterial action (known as biofouling) or wear due to the high water velocities in the system. This can lead to gradual loss in system performance, which must be mitigated by regular monitoring and maintenance.

Vacuum-accelerated consolidation methods

Methods employing vacuum assisted consolidation techniques are based on a combination of vertical drainage and pumped pore water pressure control. These methods can be used to consolidate soft alluvial soils to improve bearing capacities prior to construction and to reduce post-construction settlements to acceptable values. They provide an alternative to consolidation by the use of physical surcharge.

Surcharge methods apply total stress loading to the soil to create an excess pore water pressure which dissipates with time as the soil consolidates. Vacuum methods promote consolidation by lowering pore water pressures in the vertical drains, thereby creating an excess pore water pressure in the soil, relative to the drains.

The method requires a permeable sand blanket to be laid across the area to be treated. Vertical drains are then installed in a closely spaced grid pattern, through the sand blanket, into the soft soils. Flexible, perforated, vacuum distribution pipes are laid within the sand blanket, and the blanket is isolated from the atmosphere by a flexible low permeability membrane, sealed into a clay-filled key trench at the edge of the treatment area. The vacuum distribution pipes are connected to a vacuum pump capable of pumping both air and water.

This pump is used to create a vacuum in the sand blanket, which is communicated to the soil by the vertical drains. This means that the soil has an excess pore water pressure in relation to the vertical drains, so water will flow from the soil into the drains, with water being drawn out of the vertical drains by the vacuum pump. The drainage increases effective stress in the soil, improving soil properties. The vacuum is typically applied for a few months or longer, until an acceptably high average degree of consolidation is achieved. A schematic view of a vacuum-accelerated consolidation system is shown in Figure 6.

An efficient system can achieve and maintain a vacuum of up to 80 kPa below atmospheric pressure (Tang and Shang, 2000), which is equivalent to a physical surcharge of 4–5 m of fill. By avoiding the need for large volumes of surcharge fill, vacuum-accelerated systems can allow additional flexibility in construction programmes. Vacuum consolidation can be started during the early stages of the project. In contrast, commencement of surcharge filling may be delayed until sufficient fill becomes available. Also, on very soft soils physical surcharge mounds may have to be built up slowly to avoid an undrained bearing capacity failure. Vacuum systems do not apply undrained loadings, so the full vacuum can be applied quickly, allowing a rapid start to consolidation.

There are a number of practical problems that must be addressed if vacuum systems are to be used effectively. The most important is the minimisation of air leaks into the system, to maximise the vacuum. This requires careful construction and sealing of the membrane, and ensuring subsequently that it is not accidentally punctured. One approach is to dispose of the pumped water on top of the membrane (see Figure 6). Using water as a covering medium allows small leaks to be seen, and discourages foot traffic and wildlife, reducing the risk of punctures. Ponded water on top of the membrane has the added advantage that it forms a small surcharge load in addition to the vacuum. Air losses to the system can also occur if the water table is not near the surface, allowing air to be drawn in through the unsaturated zone beneath the key trench. If the water table is deeper than the depth of the key trench, a perimeter low permeability cut-off (such as a slurry or mix-in-place wall) may be necessary, penetrating down from the key trench to the water table.

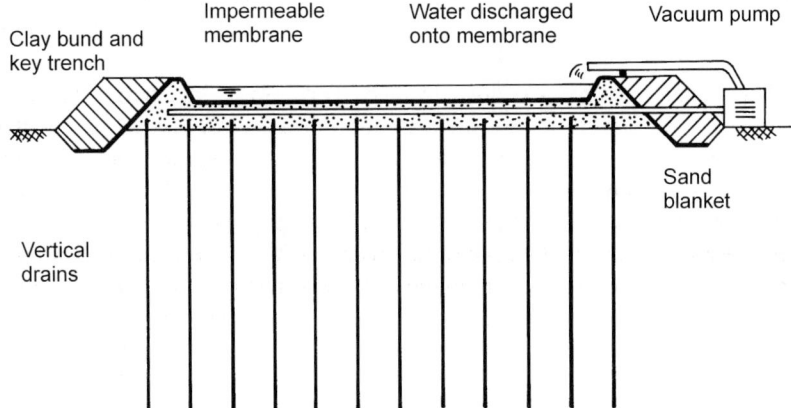

Figure 6 Vacuum-accelerated consolidation system

Vacuum consolidation methods have been used in combination with physical surcharges to achieve increased levels of consolidation. Deep horizontal drains, laid by a special trenching excavator have also been used as a means of vacuum consolidating soft soils (Anon, 1998).

A variant on the use of vacuum to promote drainage is the combination of vertical drains with dewatering (Kotera *et al.* 1991). If a significant permeable horizon exists at depth, this layer can be depressurised by pumping from deep wells or ejectors. This will create downward hydraulic gradients along vertical drains. If this type of dewatering is used in combination with application of vacuum (or physical surcharge) at the surface, 'double drainage' conditions can be generated. This allows the layer being consolidated to drain both upwards and downwards, reducing significantly the time required to achieve the target average degree of consolidation.

Electro-osmosis

Electro-osmosis is suitable for use in very low permeability soils such as clays where groundwater movement under the hydraulic gradients resulting from conventional pumping would be excessively slow. Electro-osmosis causes groundwater flow in such soils using electrical potential gradients, rather than hydraulic gradients.

A direct current is passed through the soil between an array of anodes and cathodes installed in the ground (Figure 7). The potential gradient causes positively charged ions and pore water around the soil particles to migrate from the anode to the cathode, where the small volumes of water generated can be pumped away by wellpoints or ejectors.

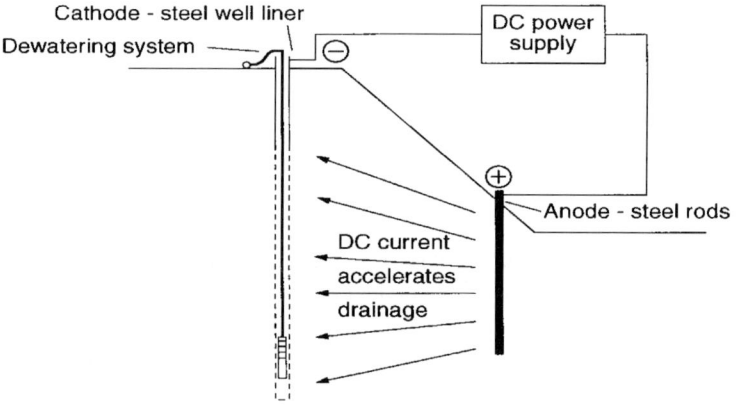

Figure 7 Principles of electro-osmosis (from Preene *et al.*, 2000: reproduced by kind permission of CIRIA)

Electro-osmosis is used rarely, mainly when very soft clays or silts are required to be increased in strength to allow stable slopes to be formed. Casagrande (1952) describes the development of the method. One of the drawbacks of the method is that it is a decelerating process, becoming slower as the moisture content decreases. In some applications, electro-osmosis is used in conjunction with electro-chemical stabilisation when chemical stabilisers are added at the anodes to permanently increase the strength of the soil.

In application the electrode arrangements are straightforward, typically being installed in lines, with a spacing of 3 to 5 m between electrodes. Anodes and cathodes are placed in the same line, in an alternating anode-cathode-anode sequence. Water is to be pumped from the cathodes, so these can be formed from steel wellpoints or steel well liners. If it is desired to use plastic well liners at the cathodes a metal bar or pipe (to form the electrode) can be installed in the sand filter around the well. The anodes are essentially metal stakes: gas pipes, steel reinforcing bars, old railway lines or scrap sheet-piles can be used.

Applied voltages are generally in the range 30 to 100 V. Effectiveness can be improved if the potential gradient can be in the same direction as the hydraulic gradient. Casagrande (1952) states that the potential gradient should not exceed 50 V/m to avoid excessive energy losses due to heating of the ground. However, it might be advantageous to operate at 100 to 200 V/m during the first few hours to give a faster build up of groundwater flow. Reduction in power consumption may be possible if the system can be operated on an intermittent basis.

Unusual applications of dewatering

Dewatering is often associated with the construction of open excavations, but there are other conditions when its application may be appropriate to improve soil properties.

- Improving soil bearing capacity. When structures or pavements are to be built on very soft soils, vacuum consolidation or electro-osmosis methods can be used to improve soil properties and reduce the potential for settlement.

- Reduction of liquefaction potential: In seismically active zones saturated loose silty soils are prone to liquefaction and loss of strength during an earthquake. Liquefaction occurs when earthquake shaking generates excess pore water pressures, which are unable to dissipate rapidly. Pumped well systems (perhaps using vacuum ejector wells) can be used to lower pore water pressures in vulnerable soils to reduce the potential for liquefaction during an earthquake. The increase in effective stress will reduce the liquefaction potential, but the major benefit accrues if pore water pressures can be lowered sufficiently to desaturate the soil. Desaturation eliminates the risk of liquefaction almost completely. The ongoing pumping requirement means that dewatering is unlikely to be used in this way to protect permanent structures, but could be used to protect slopes formed as temporary works.

- Combination with other ground treatment and foundation methods. Dewatering can be used to improve the efficiency of other methods of ground treatment. Examples include: artificial ground freezing, where dewatering has been used to reduce high natural groundwater velocities in the vicinity of freezewalls; and bored piling into water-bearing granular soils, where dewatering can improve the stability of the pile bore, reducing installation time and potentially increasing load capacity.

Conclusions

While dewatering is conventionally viewed as a means to keep excavations free from water, it also results in increased effective stress levels. Accordingly, even though nothing is added to the soil, and despite the soil not being physically re-arranged significantly, dewatering is often a *de-facto* form of ground improvement.

In addition to improving soil strength to allow stable excavation slopes and bases to be formed, dewatering techniques have also been applied to improve soil properties in other cases. Examples include pre-consolidation of soils, reduction of pore water pressures to avoid liquefaction, and dewatering used in conjunction with other ground treatment and foundation methods.

References

Anon. 1998. *Vacuum packed: a new rapid and cost effective method of pre consolidation is being used to combat the rapid settlement of soft, compressible soils in the Netherlands.* Ground Engineering, 31, (2), 18–19.

Casagrande, L. 1952. *Electro-osmotic stabilisation of soils.* Journal of the Boston Society of Civil Engineers, 39, 51–83.

Kotera, H., Sakemi, T. & Matsui, T. 1991. *A precompression method of soft ground improvement using dewatering.* Quaternary Engineering Geology (Forster, A., Culshaw, M. G., Cripps, J. C., Little, J. A.,& Moon, C. F., eds),

Preene, M., Roberts, T. O. L., Powrie, W. & Dyer, M. R. 2000. *Groundwater Control – Design and Practice.* Construction Industry Research and Information Association, CIRIA Report C515, London.

Powrie, W. & Preene, M. 1994. *Performance of ejectors in construction dewatering systems.* Proceedings of the Institution of Civil Engineers, Geotechnical Engineering, 107, 143–154.

Tang, M., & Shang, J. Q. 2000. *Vacuum preloading consolidation of Yaoqiang Airport runway.* Géotechnique, 50, No. 6, 613–623.

Kotera, H., Sakemi, T. & Matsui, T. 1991. *A precompression method of soft ground improvement using dewatering.* Quaternary Engineering Geology (Forster, A., Culshaw, M. G., Cripps, J. C., Little, J. A. & Moon, C. F., eds), Geological Society Engineering Geology Special Publication No. 7, London, 11–658.

Deep compaction of problematic soils

B.C. Slocombe
Keller Ground Engineering, Coventry, CV8 3EG

Introduction

Problematic soils can be naturally occurring, man-made or natural soils that have been displaced naturally or by man. These can give rise to many geotechnical difficulties including inadequate bearing capacity, the potential for unacceptable settlements and slope instability.

Engineers have many options available to overcome such problems. These include excavation/replacement, ground improvement, piles and other geotechnical processes. The choice will depend upon many factors including the type and size of development, programme and, probably most importantly, cost. Excavation and replacement is becoming less attractive due to the tax imposed when disposing materials off-site, particularly where the material is contaminated. Most piled schemes incorporate a system of quite heavily reinforced pile caps and ground beams, possibly with suspended floor slabs, to transfer the building loads onto the piles or pile groups. Whilst certainly very positive in terms of settlement control, piling can prove to be expensive, particularly in the case of relatively lightly loaded structures. Ground improvement by deep compaction treats soils in such a way as to improve their load bearing and settlement properties. This allows buildings to be supported on relatively simple and lightly reinforced foundations at a shallow depth below ground level, with values of total and differential settlement being within acceptable limits.

The two most widely used methods of ground improvement are Vibratory Stabilisation (referred to as "Vibro" herein) and Dynamic Compaction. The Vibro method far outnumbers that of Dynamic Compaction in terms of contracts performed each year in the UK.

Vibratory stabilisation: equipment

The stabilisation is carried out by means of depth vibrators. These are large poker vibrators 35 to 45 cm in diameter with a basic length of about 6.0 m, extension tubes being easily added to treat soils to greater depths.

Problematic Soils. Thomas Telford, London, 2001

Inside the poker the vibrations are produced by electrically or hydraulically driven eccentric weight assemblies and these produce high centrifugal forces in a horizontal plane at frequencies between 30 and 50 Hz. The amplitude of the vibrator varies between 5 and 9 mm.

The poker, in its basic form, weighs about 2 tonnes and its nose is tapered to assist penetration. Just above the vibrator tip, and in some cases along the shaft, jet nozzles emerge from the main housing. Depending on the method of stabilisation being used, compressed air or water at high pressure emerges from these jet nozzles during the Vibro operation.

The poker is generally suspended from a conventional tracked crane, and flexible leads carry the power to the vibrator from a portable generator or hydraulic powerpack, these being sometimes mounted on the back of the crane. Independent compressors or pumps boost the air or water to the required pressure before entering the vibrator. Current and voltage or hydraulic pressure can be monitored and this information provides a useful guide as to the degree of stabilisation being obtained.

A comparatively recent development for the UK market is the use of purpose made machines called Vibrocats where the vibrator is mounted on leaders and pulled into the ground by means of an hydraulic winch which is capable of developing 'pull down' forces of up to 180 kN. Not only does this 'pull down' facility speed up the treatment process but also in forming the compaction points, it is able to effect much better compaction, the effect being greatest in granular materials. The 'pull down' also obviates the need to pre-bore on many sites where the made ground is relatively dense or stiff at the surface.

This advanced Vibro technology has led to the introduction of the 'Bottom-Feed' vibrator. This is basically a standard unit with a heavy duty tube welded down one side, bending inwards at the vibrator tip to ensure a central location for the stone supply. The stone is fed down this tube via a stone reservoir and a reception chamber at the top of the 'Bottom-Feed' vibrator assembly. The whole system is charged with stone by means of a skip which travels up and down the leaders. There is a 'bell' valve between the main stone reservoir and the reception chamber and a supply of compressed air can be applied to the stone supply system as necessary. Virtually fully automated-construction computer-programmed Vibrocat rig units are currently operational in Germany.

These purpose-built bottom-feed machines achieve significantly higher production than the conventional crane-hung units. There is an increasing trend to convert all Vibro equipment to purpose-built rigs, with the added advantage of computer-generated construction records for every stone column.

Techniques
The Specification for Ground Treatment using Vibro Stone Columns published by the Building Research Establishment (2000) defines Vibratory Stabilisation techniques as:

Vibro compaction

"In loose permeable granular soils the vibratory action of the poker will densify material immediately surrounding the column, thus significantly improving its geotechnical properties. Stone is sometimes added in instances where silt and/or clay is present in sufficient quantity to inhibit densification by vibration or where enhanced drainage in addition to compaction may be required. Stone is also sometimes added to produce superior densification and speed up the process of construction. Further details are given elsewhere (Slocombe, 2000).

Vibro stone columns

"Formation of the stone columns should provide a composite ground structure of in-situ material and stone columns acting as vertical reinforcement, which overall has lower compressibility and increased bearing capacity. The Vibro "wet" method uses a water flush whilst the Vibro dry methods use an air flush system. Stone fed from ground level is called the Top-Feed technique.

Vibratory Stabilisation is rarely performed in the UK without the addition of stone.

Method

Wet methods

The Vibro Compaction and Vibro Stone Column Top-Feed 'wet' techniques utilise water-flushing to assist the insertion of the vibrator to the design depth whilst the presence of water under pressure in the annulus surrounding the vibrator avoids collapse of the hole.

On reaching the design depth with Vibro Compaction, the vibrator is lifted in stages to the surface with holding periods at each level. It is essential to maintain contact between vibrator and soil throughout the lifting operation and it is for this reason that the addition of coarse stone provides superior densification and faster construction, albeit at additional cost. In the classical sand compaction, site material is used and the increased density achieved results in a lowering of the site surface.

With the Vibro Stone Column wet method, the water is used to maintain stability of the sides of the hole, even in very soft clays, silts or water bearing fine-grained sands which are liable to slump, for a long enough period to enable stone to be fed to the base of the hole. The high pressure jetting can also be used to flush out large diameter in particularly soft or unstable soils such as thin peat layers.

On reaching the design depth, the vibrator is maintained in the hole and a charge of stone allowed to filter down the bore to form charge of stone. The vibrator is reintroduced and compacts the charge of stone (and surrounding soils if granular). The process is repeated to form a dense stone column, tightly interlocked with the surrounding ground, to the surface.

The water supply for the wet process needs to be of the order of 3000 to 5000 gallons per rig per hour. Furthermore, suitable drainage facilities have to be provided, normally ditches and settling ponds, usually by the Main Contractor and there are sometimes difficulties in getting the effluent accepted by drainage authorities. Site surfaces tend also to become heavily contaminated with silt fines and it is sometimes necessary to remove and replace surface hardcore beneath building and road areas.

Dry methods

The Vibro Top-Feed and Vibrocat Bottom-Feed dry methods both use compressed air to assist penetration and prevent suction of the ground as the vibrator is withdrawn, particularly in clay soils.

The Vibro Top-Feed dry method is adopted in all soils where the hole formed to the design depth by the vibrator remains stable when the vibrator is withdrawn. It is sometimes possible to enhance the stability by lining the hole with stone during penetration but this procedure appears to be more applicable to the higher frequency vibrators than the lower. A charge of stone is then introduced and a dense stone column formed to the surface as before.

The 'Bottom-Feed' method is a completely dry process and has been designed specifically to avoid the use of high volume water-flushing on sites with a high water table or underlain by weak clays, silts etc which would normally slump and require the use of the wet system. The main feature of the method is that the vibrator remains in the ground during the whole construction process to provide the necessary support to the hole, stone being supplied directly to the base via the delivery tube, whilst the pull down facility ensures stone columns of very high integrity.

The grading of the stone backfilling is normally within the range of 100 to 40 mm for the Vibro Top-Feed dry method and 40 – 20 mm for the Bottom-Feed and "wet" techniques. The dry methods dominate the UK market since they permit subsequent construction operations to follow close behind the Vibro works.

How the stone columns work

The action of the stone columns (sometimes referred to as compactions or pressure points) is dependent on the soil type. In granular soils, i.e. sands, sand and gravel, ash, brick rubble, in addition to a column of very compact stone being formed in the ground, the action of the high frequency vibrations also increases the density or state of compaction of the granular soils between the stone columns.

However, with cohesive soils, i.e., clays, these are little affected by vibration and the formation of the stone columns does not appreciably alter the shear strength of the clay in the short term. The clay does, however, confine the stone column and this passive resistance allows the column by virtue of its dense granular nature to develop a high bearing capacity relative to the surrounding ground. In this way an overall increase in bearing capacity is obtained and at the same time the stiffness of the stone columns reinforces the stabilised clay substantially reducing the amount of consolidation settlement which can occur.

Many soils are of course, a mixture of these two extreme soil types. In spite of this, the formation of the stone columns at regular intervals beneath the applied load compacts and strengthens the soil mass, and careful choice of treatment depth and stone column spacing generally ensures that total and differential settlements are kept within acceptable limits.

Vibro theory

The simplistic approach is that Vibro stone columns are relatively rigid springs, the rigidity of which depends on the confining action of the soil. Where the soils are of a clean granular nature, they will be enhanced by densification effects due to the poker vibrations, thus providing even greater column rigidity. Clayey soils can also be subjected to lateral pre-loading during proper stone column construction, but this is generally ignored. The applied loads are then supported by a system of relatively stronger springs surrounded by weaker springs that represent the soil confining the stone columns.

There are three main design procedures in use in the UK:

<ul>
<li>a)   Hughes and Withers (1974)</li>
<li>b)   Baumann and Bauer (1974)</li>
<li>c)   Priebe (1976, 1988 and 1995)</li>
</ul>

These have been commented elsewhere (Raison, 2000), and comments are summarized below.

When considering the bearing capacity of columns using Hughes and Withers, the key parameters are:

K_{pc} = passive earth pressure coefficient for the column
K_s = assumed to lie between K_0 and K_p and is usually taken as 1 for clays and K_p for sands
Φ' = friction angle for the stone column
c_u = undrained shear strength of soil
σ_v = effective vertical stress in soil.

The Baumann and Bauer (1974) and Priebe (1976, 1988 and 1995) methods are then used to assess the reduction in settlement as a result of the construction of the Vibro Stone Columns. The key parameters for Bauman and Bauer are:

K_s = assumed to lie between K_0 and K_p
K_c = assumed to lie between K_a and K_0 but usually taken as K_0
E_s / E_c = ratio of stiffness of soil to column

The key parameters for Priebe (1976, 1988 and 1995) are:

K_{ac} = active earth pressure coefficient for column
K_0 = assumed to be 1 for all soils
Φ' = friction angle for stone column
E_c / E_s = ratio of stiffness of column to soil
v_s = Poisson's Ratio for the soil

There is some debate on the merits of the two methods. One problem with the Baumann and Bauer approach is that the usually quoted values of K_s do not follow the normally accepted trend of lower values with finer soil. This results in the Baumann and Bauer method predicting a settlement improvement ratio of about 40% higher in clay soils than the Priebe method (see Appendix). It was stated by several attendees at the above meeting that the Priebe approach was considered to be probably the more realistic method for clayey soils. There is also increasing evidence to suggest that the Priebe method can be applied to both primary and secondary settlements.

Further design can be applied by the fact that stone columns act as drains to effectively accelerate the rate at which the reduced settlements will occur. This is particularly useful when considering the support of road embankments or where a site is upfilled for a period of time before the structure or floor slabs are constructed.

Similarly, stone columns are inclusions of material with high angle of friction that can be used to enhance the stability of slopes, particularly for roads embankments.

Design of vibratory stabilisation schemes

Most Vibro stone column schemes are carried out on the basis of treating throughout the full depth of the weak soils terminating after a nominal penetration of better bearing strata. The pressure points are specifically located beneath bases and footings with the number and spacing being dependent on the soil conditions, foundation loadings and settlement tolerances. This full depth treatment approach has the major advantage that the ground is effectively investigated at close grid centres by the penetration of the vibrator and reaction of the soil to the construction of the stone columns. Any local anomalies can therefore be investigated and the design adjusted to accommodate.

However, where there is a substantial thickness of weak deposits, i.e., in excess of about 6.0 m it is either not economical or impracticable to treat the full depth of the weak soils. In such cases treatment is carried out to such a depth as to ensure that the majority of the stresses from the proposed buildings are accommodated within the treatment zone and that the stresses imposed on the untreated materials beneath the treated zone are of a low order and unlikely to contribute significantly towards total and differential settlement. This approach, however, is not often feasible for extensive deposits of weak alluvial clays and silts. Clearly, if this approach is adopted it is imperative that sufficient site investigation information be available to allow the settlement of the untreated soils at depth to be predicted with reasonable accuracy and confidence. The history of the site obtained by desk studies is also very useful.

When dealing with made ground the limited depth or strengthened crust approach does involve a degree of risk as fill material is inherently variable and may contain undesirable constituents between boreholes. It must be stressed that the limited depth treatment approach has in many cases been the only way of developing deep filled sites for housing and light to medium industrial purposes at reasonable cost; the cost of piling and full depth treatment for the same schemes being prohibitive.

With any scheme involving limited depth treatment, particularly on made ground, the responsibility of the various parties in accepting this design approach must be clearly understood, and the site investigation must be sufficiently comprehensive to allow the risks involved to be evaluated properly.

Control testing
Short term plate load tests
These are carried out by excavating a shallow pit and carefully bedding down a 600 mm plate on a stone column. The load is applied in increments by jacking against the underside of the base machine used to carry out the work. This test is usually complete within two hours.

Because of the depth of stressing and its short duration, this test has limited value in terms of predicting the settlement of the working foundations. However, because thousands of such tests have been performed by Vibro contractors over the years in a wide range of soil types, the short term test is a reliable form of quality control.

In granular soils these plate load tests may also be carried out between stone columns. With clay soils load tests between stone columns are not appropriate because, as already stated, the vibrations will not result in improvement of clay type soils between the stone columns.

Long term zone loading tests
These are dummy foundations of a similar size to those being used in practice, which are loaded, either directly or by jacking against kentledge. Settlements can be measured by taking levels and/or reading dial gauges. These tests are not cheap.

Standard or dynamic penetration testing (DPT) and Dutch cone testing (DCT)

These tests are very useful in assessing improvement of granular soils, but have little value when dealing with the treatment of clayey soils. It is often difficult to interpret DPT tests since no sample is recovered. For example, the DPT blow-count is the same in a loose sand as in a stiff clay.

Where vibratory stabilisation cannot be used

(a)   Sites where the made ground contains excessive quantities of degradable domestic refuse – this would decay with time and cause loss of support to the stone columns and hence settlement. Generally where the degradable refuse content is less that about 10% and fairly evenly distributed, sites can be considered for treatment but the design approach has to be conservative and whether or not Vibro is used is dependent on the building, its function, and also the client.

(b)   Recent clay fills – this is a situation where self weight movement will not be complete and whilst the introduction of stone columns may well improve matters it is still very difficult to control self weight movements. Significant depths of clay fill material should not be considered suitable for treatment if it is less than 5 to 8 years old.

(c)   A variant of the backfilled clay site is the backfilled opencast mine working, where there is the added problem caused by the restoration of the ground water table, the timing of which is totally unpredictable. When the ground water does re-establish itself there is a breakdown of the clay fill and further settlement, which can be in excess of the self weight movements.

(d)   Very thick peat deposits. Thin peat layers can be accommodated but their thickness and depth have to be considered very carefully in relation to the size of foundation and its loading. On some sites covered by up to 2.0m of peat, the peat has been excavated along the lines of the foundation over a 2.0 to 3.0m width and replaced with granular fill, Vibratory Stabilisation then being carried out in the normal way.

Design of foundations after vibratory stabilisation

With regard to reinforcement in foundations and floor slabs supported on treated ground, there is generally no need to design for spanning between compaction points. For uniformly distributed loads all that is necessary is light mesh. Clearly where foundation elements are subject to overturning moments and variable loadings, increased reinforcing is required; but in any event the treated ground can be considered as being capable of developing uniform bearing pressure.

The bearing pressures to be used after treatment are dependent on the engineering properties of the original soils before treatment. The table 1 provides a guide as to what can be achieved in terms of load bearing and settlement after Vibratory Stabilisation.

Soil Type	Bearing Pressure	Settlement Range after Treatment
Made Ground – Mixed and Cohesive	100 to 165 kN/m²	5 to 25mm
Made Ground – Granular	100 to 215 kN/m²	5 to 20mm
Natural Sands or Sands and Gravels	165 to 500 kN/m²	5 to 25mm
Soft Alluvial Clays	50 to 100 kN/m²	15 to 75mm

Table 1 Improvements achievable in terms of load bearing and settlement after vibratory stabilization

Because the stone columns are formed by working against the overburden pressure, the stone at the top of the compaction points is not as compact as at depth. It is for this reason that the main load bearing foundations should be placed at a minimum depth of 600mm below the level which treatment was carried out in order to fully develop the performance provided by the Vibratory Stabilisation contractor.

With floor areas, the treated ground is normally simply rolled and regraded as necessary prior to construction.

Supervision and control of vibratory stabilisation
Design
The most important factors in evaluating Vibratory Stabilisation schemes are the site investigation information, the number of pressure points and the depth(s) of treatment. Each of these will now be discussed in turn as follows:-

(a) The site investigation should be as comprehensive as possible leaving no areas of serious doubt concerning the nature of the soils, their engineering and chemical properties and the ground water conditions. The soils should have been proved to such a depth as to allow settlements to be calculated with confidence. The Vibro contractors should confirm that the soils are suitable for treatment using their plant and technique, and they should all be given the opportunity of examining trial pits to confirm this. Most responsible Vibro contractors will in fact conduct their own trial pit investigations at some stage either prior to tendering or before accepting an order.

(b) The number of pressure points on any site is generally determined by the soil parameters, the building loads and settlement criteria; sensible treatment depths being an essential pre-requisite.

In natural soils the pressure point numbers can be determined with reasonable confidence by calculation using the various soil parameters. However, until recently with made ground, because of the inherent variability, the maximum spacing of pressure points beneath main load bearing foundations were not allowed to exceed 2.0m whilst beneath floor slabs there was a recommended maximum spacing of about 2.5 – 3.0m. This practice had been evolved over the years by the main Vibro contractors after separate and joint discussion with many of the major Local Authorities and consultants in the UK. This maximum spacing approach had proved a very effective form of control and there were very few problems indeed.

(c) For housing, full depth treatment to shallow depth of weak soils i.e. up to about 4.0m beneath the main load bearing foundations should be considered essential. Any attempt to pull the treatment back to say 3.0m or even 2.0m below ground level is increasing the risk of foundation problems beneath the houses, particularly where the soils concerned are wholly made ground.

With industrial type buildings, whilst full depth treatment beneath structure and floor is desirable, limited depth treatment is sometimes applied to at least the floor slab. If it is also decided to limit treatment depth beneath stanchion bases it must be remembered that heavy floor loadings could induce excessive settlement beneath the stanchions.

(d) For deep filled sites or sites covered with large thicknesses of weak natural soils the philosophy of the Vibratory Stabilisation approach in these soils conditions has already been discussed. However, the main 'economy' in design is on treatment depth and the Engineer should endeavour to relate the chosen treatment depths to the thicknesses of soils being left untreated (which could affect the building if there are any unknown factors) and the load bearing and settlement requirements of the building concerned. This approach is often referred to as the formation of a "stiffened crust" within which the majority of the foundation stresses would be dissipated and is normally restricted to soils of reasonable integrity or those that respond to densification.

(e) A fairly common soil profile is where a moderate thickness of reasonably competent fill material is underlain by a relatively thin layer of highly compressible soft organic clay before encountering competent soils at depth. In this situation, there is often far more settlement potential in the organic clay than the made ground even before treatment. In some cases the treatment has been restricted to the made ground and the highly compressible organic clay has been left untreated and the building put at greater risk, with everyone believing the main problem to be the made ground. With this soil profile it is strongly recommended that these "critical" layers be carefully examined.

One very real advantage of the addition of stone when dealing with critical layers is that the forces involved in the construction of the columns normally result in larger column diameters to compensate for the lesser confining action of the weaker zones.

On-site
(a) It is common fallacy that all construction methods and standards are the same for all Vibro contactors. It is therefore suggested that:

1. Obtain from the Vibro contractors an indication of what the power criteria (current readings or hydraulic pressure readings) are for the required level of compaction. The actual readings should be recorded on the work sheets.

2. The depths of all pressure points should be recorded together with details of obstructions, any dense or soft pockets etc.

3. With the Top-Feed "dry" process ensure that the stone is added to the probe hole in relatively small charges, each one being compacted. The only exception to this is where a probe hole may be fully charged with the vibrator then being driven right through this stone to the base of the probe before forming the stone columns in the normal way.

4. The quality of the stone aggregate should be confirmed by the provision of suitable testing. The most useful is the Ten Percent Fines test, BS 812, part 111, 1992. The test samples should be soaked for when Vibro treatment is taken below the water table. The stone consumption for each column should be recorded on the work sheets.

(b) Plate load tests should be carried out in areas where there would appear to have been difficulty in forming the stone columns – this would be the result of observation on the Engineers part.

(c)   The Engineer may wish to check stone column locations. For granular soils the stone columns can be up to 200mm out of position without affecting the treatment but in clays the tolerances are smaller with the distance out of position being no more than 150mm.

To summarise, with good supervision and control, Vibratory Stabilisation offers an economic and speedy solution to many sites underlain by problematic soils, usually with considerable savings in time and materials, compared with alternative foundation schemes.

Dynamic compaction

Dynamic Compaction is the improvement of weak soils by <u>controlled</u> high energy tamping. Whereas piles make a hole in the ground which is then filled with concrete and Vibratory Stabilisation normally fills a hole with stone, the Dynamic Compaction technique does not normally add any external material. It is convenient therefore to consider the technique as dynamically pre-loading the ground at a much faster rate than can be achieved by static methods.

The prime advantage of Dynamic Compaction is that it induces improvement to substantial depths. In very uniform deposits of loose dry ash improvements have been recorded at depths of up to 20m, although 8-10m is more normal with full-scale and 5-6m with conventional equipment. This is often achieved at considerable economy compared to Vibratory Stabilisation.

Equipment

Treatment is generally carried out by dropping large weights of up to 25 tonnes in virtual free-fall from heights of up to 25m using standard crawler cranes. For the majority of the cranes available in the UK, the Health and Safety Executive regulations restrict weights to about 16 tonnes and height to about 18m. The tamping weights are normally constructed of steel box filled with concrete or thick steel plates bolted or welded together. The 15-16 tonne equipment is sometimes referred to as "full-scale dynamic compaction" whilst weights of 6-10 tonnes tend to be called "conventional" Dynamic Compaction.

Working surfaces

Because the tamping process uses very heavy cranes (usually 60-100 tonnes) most sites must be provided with a good working surface. A working carpet not only provides a stable support for the crane during the tamping operations but also provides a source of material with which to backfill the holes or imprints formed by the tamping process. The thickness of the working carpet depends on the ground conditions (and also possibly the depth of the water table) but it should be a minimum of 300mm thick and on some sites it may well need to be in excess of 1.0m. It is usually necessary to import a free-draining, preferably coarse, granular working carpet when performing treatment in the British climate where the surface soils do not meet the above requirements.

The provision of a working carpet also serves to offset one effect of Dynamic Compaction which is to reduce the general level of the site. However, when Dynamic Compaction is contemplated it is prudent for the Engineer to retain as much flexibility as possible on finished levels as the reduction in level induced by the treatment is very difficult to predict accurately.

Method of performance

Granular soils

In granular materials, i.e. sand, gravel, ash, brick rubble, etc, it is very easy to see how tamping improves engineering properties. The voids ratio is reduced, the relative density increased and the treated soil has improved load bearing and settlement characteristics.

With granular soils the treatment is usually carried out in two or three passes, sometimes combined into one; the number of passes depending on the thickness of the weak materials, the required engineering performance and position of the water table.

Whilst the soil conditions will have been very carefully assessed at tender stage, the initial period of some contracts is sometimes spent conducting trials to establish the response of the ground to tamping. The ground response, or shape tests as they are sometimes called, determines the number of drops to be employed at the first pass, carefully recording the dimensions of each imprint on completion, before backfilling with material from the working surface. The effect of the tamping is to form a cone of densified material beneath each imprint position, these cones of influence intersecting at depth; the ground therefore being improved from depth.

If a second pass is required this is normally performed by dropping the weight a lesser number of times from a lower height. The result is that the cones of influence from the second pass intersect with those of the first raising the level below which improvement has taken place.

The final or 'ironing' pass consists of dropping the weight over the whole treatment area from a reduced height (5 to 8m) and only a small number of times. This compacts the weak zones in between the first and second pass prints and also the loose granular material within those imprints

Granular soils respond very quickly to the tamping process. The minimum size of treatment area is the order of 3000 to 5000m^2, the treatment areas usually being the plan areas of buildings or groups of buildings where close together plus a two or three metre margin all round.

Clay type soils

Whilst the basic approach is similar to granular soils there is one very important difference. To illustrate this it is appropriate to refer to elementary soil mechanics theory.

The settlement of foundations on clay soils is caused by the pressures from the building loads expelling porewater from between the clay particles. Apart from the reduction in volume (representing settlement), the loss of water also results in an increase in strength over a long period of time. What dynamic compaction dramatically speed up this process.

When the weight drops onto clay type soils very high instantaneous porewater pressures are set up and these cause partial liquefaction of the clay. Preferential drainage paths are created and these allow the porewater to be expelled very much more rapidly than would be the case with static loading; thus effecting an improvement in engineering properties.

It cannot be emphasised too strongly that this requires experience and very close control on site. Excessive tamping can lead to a noticeable reduction in the strength of the clays. In some cases drainage measures using Vibro stone column, Band Drains or trenches may be needed to assist dissipation of excess pore pressures. Loose-placed clay fills tend to be relatively voided to not normally require such measures.

The rate at which the porewater pressures dissipate governs the size of sites which can be treated. Clearly it is uneconomic to have the tamping equipment standing idle at the completion of say the first pass waiting for the porewater pressures to dissipate at the beginning of the second pass. Because of this, the minimum size of the treatment area with clay type sites is generally in the order of 10 000 m^2.

The formation of every imprint in clay sites should be observed by experienced personnel. If excessive heave starts to occur after only a small number of drops, it is essential that the tamping at that print position be stopped. This may only occur over a small area with better ground elsewhere. In soft areas it may well be 7 or 8 light passes are required to achieve the desired results, whereas 4 passes are sufficient over the remainder of the site.

How dynamic compaction works

In granular soils, the improvement is effected by essentially vibratory means. In cohesive and mixed soils, as noted earlier, the technique can be considered as a method of pre-loading the ground. The key to success is constant experienced observation of the tamping work and fastidious record keeping.

Design of dynamic compaction schemes

The design approach is very similar to that of Vibratory Stabilisation whereby the depth of treatment is a function of the size of equipment adopted and the degree of improvement based on the specialist contractor's experience. Detailed knowledge of the soil constituents is therefore essential.

Control testing

Similar testing to that of Vibratory Stabilisation can be used although a number of contracts have simply involved the measurement of imprints and monitoring of site levels. In-situ tests are sometimes performed and since the technique provides treatment to large areas very quickly, the speed at which such tests provide the necessary information is important. In clayey soils, as with the performance of the treatment, it is essential that sufficient recovery period be allowed in order to avoid ambiguous results.

Standard penetration tests

This is probably the most useful in-situ test as it applicable to both granular and cohesive soils. However, being of a dynamic nature it is particularly sensitive to the presence of residual pore water pressures, quickly liquefying the stratum being tested, and producing lower than expected results. A sample is normally recovered which should be examined by the specialist contractor and the speed of provision of information is only limited by the competence of the driller. The main drawback is that a considerable amount of chiselling is often required to penetrate the very dense surface layers normally provided by the treatment.

Dynamic penetration tests

These have similar drawbacks with the added limitation of not recovering samples.

Dutch cone tests

The emphasis here is for the well established types of cone test, typified by Fugro, and not the non-standard probe type. These are ideally suited to the testing of sands because they also illustrate the soil type by means of the friction ratio. They are less successful in treated clays and are of little use for coarse fill materials after treatment.

Zone loading tests

These again are the most meaningful tests. Dynamic Compaction is probably the only economic way of treating large sites containing massive obstructions. In-situ testing is therefore very difficult and often the only valid testing is large scale zone testing on completion of the work.

Sites where dynamic compaction is not suitable or its use should be considered with suspicion

(a) Deep deposits of soft alluvial clays and silts – various attempts have been made to treat these soils using Dynamic Compaction and real success has been negligible. Vibro Stone Columns are more normally used in such situation, sometimes in conjunction with Dynamic Compaction. The treatment of any clay site should be considered very carefully, particularly if the work is to be carried out in winter.

(b) Sites with fill materials heavily contaminated with degradable domestic refuse. The technique has been used on such sites and whilst effecting some improvement, the settlement criteria for the structures erected after treatment have had to be radically changed. Treatment is normally restricted in this situation to road and parking areas where the method collapses voids and compacts inert constituents into a denser matrix within the treatment zone. The technique should not be considered as a method of overcoming long-term decay of degradable constituents.

Design of foundations after dynamic compaction

Again, this is very similar to Vibratory Stabilisation but with the added advantage of greater flexibility to locate foundations anywhere within the treatment area.

Supervision and control of dynamic compaction

Design

This is generally in the hands of the Specialist who will almost certainly be drawing upon past experience. What needs to be examined is the total energy input per square metre. This is the sum of the number of drops x the height of drop x the weight for all the passes divided by the treatment area. For normal engineering performance i.e. bearing pressure of between 100 and 200 kN/m^2 with acceptable values of settlement the energy input/m^2 is normally in the range of 50 to 150 tonne metres/m^2, irrespective of soil type. Clearly a high performance after treatment will require a higher energy input.

Factors which can radically affect the energy input to achieve the required performance of the treated ground are high water tables and large obstructions in the soils to be treated. In both cases energy is dissipated at the surface and the treatment is far more difficult to alter at depth.

With regard to the effective depth of treatment this is normally accepted as being equal to half the square root of the product of the weight x the maximum drop height. Treatment depth is particularly important when dealing with large heavily loaded areas such as steel works etc.

On site

Most of the site controls have already been dealt with in the section describing the Dynamic Compaction System. The key factor is observation and very careful record keeping. Care should be taken to ensure that the energy levels are as promised by the contractor.

Vibrations

This is a major factor in the choice between Vibratory Stabilisation and Dynamic Compaction. Whilst the former is often performed as close as 2.0 to 2.5m from existing structures and services in good condition, it is not desirable for such features to be within about 30 to 50m of the conventional and full-scale tamping operations respectively. Larger clearances are often necessary when the treatment layer or underlying stratum at shallow depth is sand, gravel or rock as these soils tend to transmit vibrations with comparatively little attenuation.

In addition to the magnitude of the vibrations, the typical frequency of Dynamic Compaction is 5 to 15 Hz which is more damaging to structures and more noticeable to humans than the Vibratory Stabilisation frequency of 50 Hz. This is recognised by the major computer companies who require more stringent vibration tolerances for frequencies below 14 Hz than above. The effect of tamping vibrations can be monitored with special equipment and cut-off trenches can sometimes be used to protect adjoining features.

The human being is particularly sensitive in detecting vibrations and amplitude is the parameter normally considered. Again, Dynamic Compaction produces the higher amplitude and even at 50m clearance would be classified as "perceptible". The equipment is highly visible and a psychological reaction that damage is being caused, even when values for far below any threshold level, can occur.

Safety is a prime consideration and must be carefully examined when working near roads, railways and other operations on site. It is not unknown for material to be ejected 50 to 60m on impact of the weight particularly when the site surface is wet although a coarse working carpet tends to reduce the problem. This can cause problems in the phasing of development of a site but can be overcome by using Vibratory Stabilisation for successive visits.

Summarising Dynamic Compaction is a very economic and technically sound method of treating larger sites provided the above constraints are recognised.

Choice of ground improvement techniques

The choice between Vibratory Stabilisation and Dynamic Compaction depends on the soils conditions, size of the site and environmental conditions. The presence of near surface groundwater, soft cohesive soils and proximity of structures, services and roads place severe restrictions on the use of Dynamic Compaction and in these cases Vibro methods are normally adopted. On large sites Vibratory Stabilisation has been used around sensitive boundaries with Dynamic Compaction being used for the main area.

Closing comments on the use of ground improvement
Techniques

There is a very wide and expanding range of techniques available to the Engineer for the solution of today's foundation problems and these processes are generally performed by very experienced and skilled contractors. In addition to conventional improved bearing capacity and settlement control both methods can be used to overcome liquefaction problems under seismic conditions. In this situation, the Vibratory Stabilisation has the added advantage of stone columns also providing enhanced drainage characteristics. Stone columns can be used to improve slope stability in certain situations and have also been used to improve the stability of weak soils in a tunnelling project. The essence to all the above is that improved ground conditions are achieved by these two methods.

It will be appreciated that, apart from the structures, the key factor with any development is the soil profile. The performance of ground improvement methods is aimed at achieving a suitable engineering solution to the particular problem. The economic design of such schemes can only be achieved by the performance of appropriate site investigation.

References

Bauman, V., & Bauer, G.E.A. 1974. *The performance of foundations on various soils stabilized by the vibro-compaction method.* Canad. Geotech. Journal 11, 4, 509-530.

BS812 part 111. 1992. *Testing aggregates. Methods for determination of ten per cent fines values.* Code of practice. British Standards Institution, HMSO, London.

Hughes, J.M.O., & Withers, N.J. 1974. *Reinforcing of soft cohesive soils with stone columns.* Ground Engineering. Foundation Publications Limited, May, 42-49.

Watts, K. 2000. *Specifying vibro stone columns.* BRE report 391.

Priebe, H.J. 1976. *Abschatzung des Setzungsverhaltens eines durch Stopfverdichtung verbesserten Baugrundes*, Die bautechnik 53, H.5.

Priebe, H.J. 1988. *On assessing the behavior of a soil improved by vibro-replacement.* Bautechnik 65 (1), 23-26.

Priebe, H.J. 1995. *The design of vibro replacement. Reprinted from: Ground Engineering*, Technical paper 12-61 E.

Raison, C.A. 2000. *Input parameters for the design of vibro stone columns.* Presentation at the launch of the CIRIA Research Project, RP604.

APPENDIX: Comparison of Baumann & Bauer and Priebe Vibro design methods.

Baumann & Bauer Pad footing	Soil Clay	Soil Sand	Priebe	Soil Clay	Soil Sand	
Stone column diameter	500	600		500	600	mm
Average vertical effective stress	-	-		15	40	kPa
Initial soil friction angle phi'	0	35.0		0	37.5	Deg
Final soil friction phi'	0	37.5		0	38.8	Deg
Soil shear strength cu	50	0		50	0	kPa
Soil Young's Modulus Es	7.5	25		7.5	25	MPa
Soil Poisson's Ratio v	0.5	0.2		0.2	0.2	
Earth pressure coefficient Ka	1	0.27	Kc=Ka	0.22	0.22	
Earth pressure coefficient Kp	1	3.69		-	-	
Earth pressure coefficient Ks	1.25	0.85		-	-	
Earth pressure coefficient Kc=Ko	0.36	0.36		0.36	0.36	
Applied foundation load q	150	150		150	150	kPa
Unit area per column A	1.25	1.25		1.25	1.25	m^2
Equivalent radius a	0.63	0.63		-	-	m
Stone column friction angle	40	40		40	40	Deg
Stone column Young's Modulus Ec	40	40		40	40	MPa
Stone column spacing	1.2	1.2		1.2	1.2	m
Stone column area Ac	0.196	0.283		0.196	0.283	m^2
Area ratio Ac/A	0.16	0.23		0.16	0.23	
Ratio A/Ac	6.37	4.42		6.37	4.42	
Ratio Ec/Es	5.33	1.6		5.33	1.6	
Ratio Es/Ec	0.19	0.63		-	-	
Pc/Ps	11.57	5.39		7.18	7.67	
Pc	652.3	405.8		546.5	458.6	kPa
Ps	56.4	75.2		76.1	59.8	kPa
Basic improvement factor n_0	2.66	1.99		1.97	2.51	

Enhanced cement stabilisation of contaminated clay soils

A. Maries[1]&[2], *C.D. Hills*[2], *K. Whitehead*[2] *and*
C. L. MacLeod[2]
1) Cementation Skanska, Bentley House, Jossey Lane, Doncater,
South Yorkshire, DN5 9ED
2) Centre for Contaminated Land Remediation, Centre for
Contaminated Land Remediation, Natural Resources Institute,
University of Greenwich, Chatham, Kent ME4 4TB.

Introduction

The ratification of the EU Directive on Landfills seems likely to ensure a concerted European effort to deal with hazardous waste management, including pre-treatment prior to land filling. The most common technology for such pre-treatment is solidification/stabilisation in which hazardous species of a waste are controlled or immobilised, either by chemical bonding or physical entrapment by the addition of a cementitious additive (Sweeney *et al.*, 1998).

To date, no one solidification process has been developed which is optimal for the most common types of hazardous wastes. However, cement-based solidification/stabilisation can immobilise a variety of wastes including both organic and inorganic materials. The binders commonly used include Portland cement, fly ash, cement kiln dust, natural pozzolans and various industrial by-products. In principle, cement based systems rely on the formation of silicate or aluminosilicate matrices in which the waste constituents are incorporated either chemically or physically, thereby limiting release of toxic components to the environment (Anderson *et al.*, 1979).

However, it has been found that the complex nature of many wastes may inhibit the reactions that are responsible for effective solidification and stabilisation. Furthermore, the integrity of cement-bound waste forms can be influenced by subsequent atmospheric carbonation, which is a slow, natural process affecting a wide range of cementitious materials over timescales of many years.

Problematic Soils. Thomas Telford, London, 2001

Accelerated carbonation

The carbonation process can be artificially accelerated to improve the performance of Portland cement-based materials. Although this is a comparatively little-known technique, patents relating to its industrial applicability extend back to the mid 19th century (Maries, 1998). The chemistry of the process has been extensively investigated, but the short and long term effects and mechanistic details of accelerated carbonation have received attention only more recently.

The accelerated carbonation reaction can be strongly exothermic and cement hardening can be extremely rapid with compressive strengths of several MPa being developed within minutes in monolithic solids. Normal hydration reactions continue after the supply of CO_2 has been terminated thus resulting in durable concrete with high dimensional stability. Previous work using accelerated carbonation to treat hazardous wastes containing high concentrations of heavy metals (Lange, 1997) has shown that unconfined compressive strengths in the order of 10 MPa can be achieved in small monolithic samples within a few minutes. Following a storage period of up to 2 years, these monolithic samples showed no sign of structural or chemical instability. This strongly suggests that the reaction products from accelerated carbonation are very durable.

Recent investigations undertaken by the Centre for Contaminated Land Remediation at the University of Greenwich have demonstrated that introducing an accelerated carbonation treatment step into a cement solidification process for remediating contaminated land can bring a number of technical and commercial benefits that include:

- Increasing the rapidity of the process (treatment is carried out in minutes);
- Potential for improved encapsulation for a range of contaminants;
- Possibility of using unusual and more cost effective cementitious binder systems;
- Minimal additional costs over traditional cement-based systems.
- Sequestration of CO_2, since as much as 50% by weight of binder can be taken up in some cement-soil mixtures.

Accelerated carbonation technology (ACT) is a process exclusively licensed to Blue Circle Land Remediation System Ltd.

Preliminary investigations

The main objective of the work described in this paper was to demonstrate the application of accelerated carbonation to heavy metal contaminated soil on a field scale, using engineered test cells which could provide for a long term monitoring programme.

The field site used in this investigation is a reclaimed salt marsh overlying London clay, in Dartford, Kent. Pyrotechnics such as fireworks and flares were manufactured there from the late 19th century through to 1989 under the ownership of several different companies, after which the University of Greenwich purchased the site. Separate firework types were manufactured in individual huts, with test and disposal areas distributed across the site resulting in distinct hotspot areas of contamination.

Site characterisation and selection of test plot

Previous characterisation of the site by the University of Greenwich had indicated that concentrations of 32% copper, 3.3% lead and 2.6% zinc were present in selected locations, with a major "hot-spot" identified towards the front of the site. It was this particular hot-spot that was selected as the field trial test plot.

A 10 m by 20 m area was characterised in detail by excavating 11 test pits down to depths of 0.84 m. The top 0.2 m of the soil profile was composed of surface rubble and a hard crust, over brown clay. London clay was found at a depth of 0.6 m and the water table at 0.8 m. Soil samples were taken throughout the soil profile, digested with aqua regia and analysed by Inductively Coupled Plasma-Atomic Emission Spectroscopy (ICP-AES).

Heavy metals were generally found to be concentrated in the upper 0.3 m of the test plot, and the range of metals determined is given in Table 1. The main contaminants of concern were copper (Cu) and zinc (Zn), although high concentrations of lead (Pb) and chromium (Cr) were also found.

		Cr	Cu	Pb	Zn
Concentration of metals in soil [mg/kg]	Test plot maximum	71	96,000	750	81,000
	Composite soil sample	10	17,500	200	64,050
Metals leached [mg/l]	Portland cement	0.061	6.325	0	0.129
	Non-carbonated EnvirOceM$^{®*}$	0.028	6.289	0.003	0.107
	Non-carbonated EnvirOceM$^{®}$ Carbonated	0.017	1.694	0	0.043

Table 1 Chemical composition of test plot soils and composite laboratory sample, and laboratory leaching results

---

* Blue Circle EnvirOceM$^{®}$ is a registered trademark owned by Blue Circle Industries plc.

Laboratory tests

A laboratory programme was carried out with the aim of determining the optimum mixture proportions needed to immobilise metals in the site soil sample using accelerated carbonation.

A 100 kg composite soil sample was created for the laboratory programme by blending 25kg of hot spot material and 75 kg of soil from an adjacent area containing background levels of contamination. For the purposes of the laboratory testing and to allow optimisation of the water to solid ratio (W/S), all soil was dried for 48 hours at 55° C and subsequently ground to pass a 2.8 mm sieve. The chemical composition of this composite sample is shown in Table 1.

A three-stage evaluation was employed comprising preliminary static screening, dynamic carbonation (with carbon dioxide uptake assessed by means of carbonate content determined thermogravimetrically), followed by performance testing (leach testing). As a result of this assessment two binders, normal Portland cement and a proprietary cement (Blue Circle EnvirOceM ®, a fine ground sulfate-resisting composition) were selected for use in the field.

Leaching tests

Whilst the degree of carbon dioxide uptake by the binder-soil system is a useful indicator of the degree of reactivity of the mix, it is also necessary to determine the effectiveness of the heavy metal immobilisation within the carbonated stabilised systems. Immobilisation performance is most commonly evaluated by means of leaching tests, which can either mimic leaching in real environments or assess leaching under harsher conditions, to accelerate the effect of environmental deterioration.

It was decided to adopt two leach test procedures for the project. All samples were screened using DIN 38414 Part IV (DIN, 1984), a simple water leach procedure which is a generic test similar to the European test methods. The more complex TCLP test (USEPA, 1996), (which uses buffered acetic acid) was applied to a selection of the mix conditions (Lewin et al., 1994; prEN, 1996). The field trial samples were also tested using both the DIN and TCLP tests.

The high clay content of the soil was initially considered to be problematic, as the accelerated carbonation process relies on the constant exposure of fresh surfaces to carbon dioxide during mixing to maximise the reaction. At high W/S, the soil became extremely cohesive and difficult to mix, leading to poor CO_2 uptake. But despite the adverse soil type, significant increases in heavy metal immobilisation were obtained after carbonation, as shown in Table 1. The range of W/S 0.2-0.3 was determined as optimum for the field trial, as higher water contents tended to cause greater leaching of heavy metals from the treated materials.

Field trial

The field trial was carried out in September 2000 at the site, to evaluate the performance of accelerated carbonation and the use of EnvirOceM® in comparison to Portland cement, and to demonstrate the feasibility of applying the process on a larger scale. Four separate test cells were constructed in order to provide a test facility that could be monitored on a long term basis:

- Cell 1: untreated soil
- Cell 2: soil mixed with a 20% w/w addition of Portland cement
- Cell 3: soil mixed with a 20% w/w addition of EnvirOceM®
- Cell 4: soil mixed with a 20% w/w addition of EnvirOceM® and then carbonated.

Test cell design

Four test cells were constructed for the field study, which each measured 5 m by 10 m and were lined with an impermeable HDPE geomembrane liner in order to prevent contamination either from adjacent cells or the surrounding groundwater. A geotextile matting was also placed underneath the liner to provide additional protection against puncturing.

Test cells were graded in depth from 0.3 m to a limit of 0.6 m (due to the high groundwater level on the test site), so as to allow natural movement of the leachate towards the drainage system, which consisted of a perforated drainage pipe and inspection chamber. The inspection chambers have a dual purpose: leachate samples can be taken for monitoring purposes and the test cells can also be drained if required to prevent flooding. Lysimeters were also installed throughout each test plot at depths of 0.1, 0.25 and 0.4 m to permit sampling of the pore water through the soil profile on a long term basis.

Treatment Procedure (cells 1, 2, 3)

Because of the cohesive nature of the soil, the excavated material was passed through a 'Rotamill' crusher several times to shred the soil and also blend the hotspot material with the surrounding low-level contaminated soil. This shredding stage was found to reduce agglomeration in subsequent stages of the treatment process. Cell 1 was filled with soil taken from this stage of the operation.

Cement binder addition was made at a 20% dosage w/w soil, and soil and binder were passed through the crusher once more to achieve thorough mixing. The materials to fill cells 2 and 3 were taken at this point, after mixing with Portland cement and EnvirOceM® respectively.

Carbonation process (cell 4)

The accelerated carbonation treatment process was applied to the contents of cell 4 as follows. EnvirOceM$^®$ and soil blended as for cells 2 and 3 was fed by conveyor in 2 tonne batches into specially designed carbonation chamber consisting of a readymix concrete truck mixer of 8 m^3 capacity specifically adapted for the field trial, with a CO_2 feed to the rear. CO_2 was added to this chamber from a pressurised liquid source, having been converted to gaseous form through two banks of four 6 kW vaporisers. A two-minute period was allowed to purge the chamber, and then the soil/binder mix was then dynamically carbonated for 20 minutes. The carbonated material was then deposited in cell 4.

The accelerated carbonation reaction was found to be considerably more exothermic than expected, and temperatures of up to 75°C were observed immediately after treatment.

Field trial results

Samples of the treated cell materials were taken from different batches throughout the trial, and subsequently leach tested using both the DIN and TCLP protocols employed in laboratory tests. Results selected from the TCLP leach tests are presented in Table 2. Levels of copper, lead, nickel and zinc are all substantially reduced in the TCLP extract from cell 4 compared to the untreated cell 1. Chromium levels are slightly higher, but still well within regulatory limits.

Cell	Contents	Cr	Cu	Ni	Pb	Zn
1	Untreated Soil	0.01	0.24	0.095	0.028	14
4	EnvirOceM$^®$ Carbonated	0.02	0.14	0.040	*bdl*	3

bdl = below detection limit

Table 2 Metals leached from treated and untreated soil from test site [mg/l] (TCLP test protocol)

Discussion

This investigation has convincingly shown that the application of accelerated carbonation can enhance the immobilisation of certain heavy metals in a heavy clay soil. Such a treatment does not introduce any extra environmental burden, since waste CO_2 can be utilised which is combined into solid compounds in the soil and thereby prevented from being released into the atmosphere. However, the level of Cr in the leachate from the carbonated soil showed a slight rise and this was unexpected. Previous work in the laboratory (Hills and MacLeod, 1999) showed that Cr was retained more effectively in carbonated systems in comparison to non-carbonated controls. The result obtained in this trial cannot be easily explained without further investigation, but may be linked to the speciation of Cr present in the carbonated EnvirOceM®-bound soil.

It is considered that the larger than expected exotherm is a consequence of applying CO_2 gas to the blended soil and cement at the same time as mixing, which constantly renews the surface available for carbonation. It may therefore be worthwhile attempting to optimise the mixing/dosing regime in order to achieve the maximum extent of reaction. Furthermore, because the carbonation exotherm (4,000 kJ/kg CO_2) generates heat at a rate an order of magnitude higher than that needed to vaporise liquid CO_2 (364 kJ/kg), a large quantity of extra thermal energy is available from the process, which could for instance be used to remove organic contamination by steam stripping.

Significant differences were observed in the appearance of the material treated by accelerated carbonation compared to that in the other cells. The final product displayed granular properties, indicating that the heat produced during this process may also achieve accelerated soil conditioning as an additional benefit, even in a fairly 'difficult' clay soil like that encountered in this trial.

Other more fundamental investigations undertaken in parallel with this field trial have shown that the carbonation reaction promotes the rapid removal of calcium from the binder, producing large quantities of solid calcium carbonate which stabilises and encapsulates contaminants in the treated product. The decalcified binder component is also polymerised in this process and develops a significant sorptive capability, thus further contributing to contaminant retention in the stabilised solidified product.

The laboratory studies had indicated that the optimum W/S for the carbonation process for the soil used was between 0.2 and 0.3. The moisture content of the stockpile material was determined on site using moisture probes previously calibrated in the laboratory, and was found to vary from 10% at the surface to 40% in the centre of the stockpile. After blending with the binder, the moisture content of the mix was typically between 20% and 30%. It was noted, however, that had the weather conditions been adverse then considerably higher moisture contents would have been observed which might have had a detrimental effect on the stabilised product.

Conclusions

Laboratory studies and the subsequent field trial have demonstrated that significant improvement of heavy metal immobilisation can be obtained by using accelerated carbonation to enhance conventional cement stabilisation, despite the carbon dioxide uptakes of a predominantly clay soil being lower than anticipated after treatment. Work is currently underway to optimise and further enhance carbon dioxide uptake, investigating alternative binder systems and mixing equipment.

The preliminary results of the field trial demonstrate that it is possible to successfully treat a heavy metal contaminated soil to retain pollutants at below levels required by current regulations, using a simple, cost-effective technology. Further full-scale trials are planned, to be carried out on a wider range of soil types and pollutants.

Acknowledgements

The authors would like to thank Blue Circle Industries, British Oxygen Company and P Forker Ltd for their support of this project by providing materials and equipment.

References

Anderson, M. A., Ham, R.K., Stegemann, R. & Stanforth R. 1979. *Toxic and Hazardous Waste Disposal*, 2, R.B.Pojasek (ed), Anne Arbour Science, MI., USA.

German Standard. 1984. *Method for the Examination of Water, Waste Water and Sludge*, (DIN 38414).

Hills, C.D., MacLeod, C.L. 1999. *Carbonation of solidified wastes-mechanistic and process considerations*. Proc. Int. Conf. Waste Stabilisation and Environment, Lyon, France

Lange, L.C. 1997. *Carbonation of cement: solidified hazardous waste*. Unpublished PhD Thesis, Queen Mary and Westfield College, London.

Lewin, K., Bradshaw, K., Blakey, N. C., Turrell, J., Hennings, S. M., Flavin, R. 1994. *Leaching tests for assessment of contaminated land*: Interim NRA guidance, NRA.

Maries, A. 1998. *Utilisation of carbon dioxide in concrete construction*, Proc. International Symposium on Sustainable Development of the Cement & Concrete Industry, CANMET/ACI, Ottawa.

PrEN 12457 Draft European Standard. 1996. *Leaching: compliance test for leaching of granular waste materials*. CEN.

Sweeney, R.E.H., Hills, C.D. & Buenfeld, N.R. 1998. *Investigation into the Carbonation of Stabilised/Solidified Synthetic Waste, Environmental Technology*, 19, 893-902.

USEPA. 1996. *Test Methods for Evaluating Solid Waste– Physical/Chemical Methods* (SW-846).

No problematic soils, only engineering solutions

S. Leroueil
Université Laval, Ste-Foy, Québec, Canada, G1K 7P4

Introduction

Are soils problematic? The author finds pleasure in examining, testing and using them. They behave in a rational manner, certainly more rationally than most people. They allow us to construct buildings, infrastructures, dykes for canalising rivers and protecting populations from floods, dams for irrigation and electricity production, amongst numerous other engineering applications. Generally speaking, they behave correctly. It is up to us to understand their behaviour and to adjust our requirements accordingly or to modify their behaviour as we do with soil improvement techniques.

After some generalities on the topic, the paper presents several cases in which problems were mostly due to a lack of understanding. The paper also indicates how a risk assessment approach can be used to minimize the consequences of potential problems.

Generalities

To introduce his thoughts on the theme of this symposium, the author will begin with three anecdotes from domains others than geotechnical engineering:

(1) It has been recently observed that there were areas of stratospheric ozone depletion and holes in the ozone layer, the planet's natural UV filter. In fact, we probably have been extremely fortunate that, thanks to recent technologies, this observation has been made. Without it, there would have been disruption in planetary biological processes with, in particular for human beings, an increase in the number of skin cancers, cataracts, etc., without an understanding of why this was happening. From this observation, a good understanding of the causes of the phenomenon may be determined and assuming a favourable political will, solutions to the problem can be found. Observation has been and will remain a key factor.

Problematic Soils. Thomas Telford, London, 2001

(2) Many diseases were problematic before Pasteur found the vaccine against rabies in 1885. Some of these diseases are no longer problematic. Smallpox, in particular, completely disappeared from the planet a few years ago. Understanding of antibodies was the key to this success.

(3) Not so long ago, months were necessary to communicate news events occurring at the other end of the world. Nowadays, because of the fantastic progress made in communication systems, we can see such events live on television. Technological development is another key for solving potential problems.

Observation, Understanding and Technological Development (OUT) may be independent yet often complementary: development and use of technologies require a good understanding of the processes involved; progress in the understanding of phenomena are often based on observation; observation and technological development can be combined, allowing, for example, extremely important progress in tunnelling techniques and control of associated surface movements.

For the author, OUT constitutes the master key to solving most engineering problems. Geotechnical engineering is not an exception. In particular, the author does not think that there are problematic soils, simply because soils behave in a rational manner. He is also convinced that, most of the time, there are engineering solutions that can be obtained through an OUT approach. There may be, however, problematic understanding of soil behaviour, problematic engineers, problematic designs, problematic specifications, problematic use of technologies, problematic situations, a problematic shortage of money, a problematic lack of political will, etc. As problematic engineers, designs, specifications or use of technologies are generally due to a lack of understanding of the situation, "understanding" appears to be the major keyword.

In his conclusions from a paper on problematic soils, Vaughan (1999) wrote, "Many of our geotechnical materials seem unusual or "problematic" because their behaviour does not fit the classic theories of soil mechanics. Rather few of our materials do. We need to modify our basic soil mechanics to include these features in our soils which control engineering behaviour." The approach advocated by Leroueil and Vaughan (1990), Leroueil (1997) and Vaughan (1999) aims at extending the concepts of limit and critical states to include the effects of these main features that are discontinuities, anisotropy, microstructure, viscosity and, in some cases, partial saturation. This provides a relatively simple framework that is relevant for most natural soils and rocks.

Understanding of soil behaviour and of the context in which soils have to behave are extremely important in an OUT approach and will be illustrated hereafter. Observation is also essential for several reasons: it helps improve our understanding of soil behaviour; it has provided the database necessary to develop semi-empirical design methods used in practice; it is also essential to good site characterisation. Finally, technologies for improving soils, dewatering soil masses, stabilising slopes, etc. are useful tools for mitigating engineering problems. Several of these technologies are presented at this symposium.

In the following sections, several aspects of geotechnical engineering on which the author has recently worked will be presented to demonstrate how a better understanding can or could be used to minimize the consequences of potential problems. Obviously, many other examples could be reported.

Kansai International Airport

The Kansai International Airport (KIA) is a large reclamation island (511 ha) 5 km offshore in Osaka Bay, Japan. Its construction began in 1987 and was completed in 1991. The airport was inaugurated in September 1994 (Arai *et al.*, 1991; Akai and Tanaka, 1999; Akai, 2000). The water depth at that location is about 18m and load applied on the sea floor was in the order of 450kPa. The subsoil consists of 18m of soft alluvial clay over several hundreds of metres of Pleistocene clay. These latter deposits show alternating clay and relatively thin sand or gravel layers. The overconsolidation ratio of the Pleistocene clay increases with depth to reach a value of about 1.4 at a depth of 160m. Down to that level, the Pleistocene clay has a liquidity index that has a tendency to decrease the depth from about 0.7 to about 0.3. The alluvial clay layer was treated with sand drains and rapidly settled in about 6 months. The settlements observed since then are due to the compression of the Pleistocene deposits over a thickness of about 120m.

Figure 1 shows the measured and calculated settlements of the Pleistocene deposits.

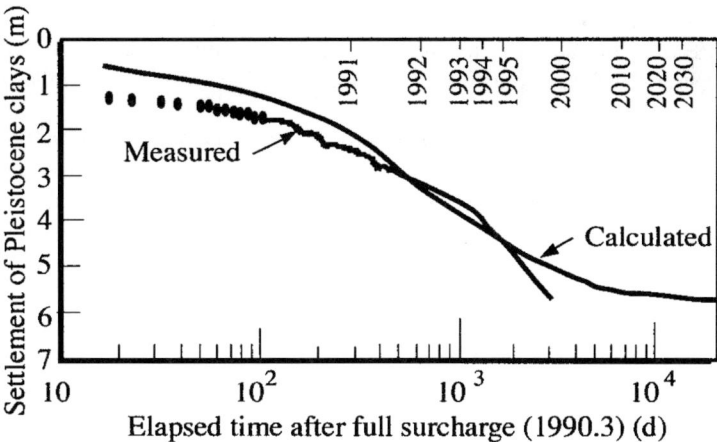

Figure 1 Calculated and measured settlement of Pleistocene clays, Kansai International Airport Island (monitoring location N° 2-1; after Akai and Tanaka, 1999)

The calculations were performed on the basis of conventional oedometer test results. At the time of the inauguration (September 1994), the measured and calculated settlements were about the same, at 4.5m. Subsequently, the rate of settlement has increased above that predicted with values of 40cm in 1995, 37cm in 1996, 33cm in 1997 and 30cm in 1998 (Akai, 2000). So in 1998, the observed settlement was close to the predicted final value, 5.84m according to Akai (2000). Akai and Tanaka (1999) and Akai (2000) presented excess pore pressure profiles measured in 1992 and 1997 (Figure 2), i.e. soon after completion of the island and 5 years later. It can be seen that the excess pore pressures decreased slowly during this period and were still very high in 1997, exceeding 200kPa between depths of 90 and 120m. The Pleistocene clay deposit is thus still far from the end-of-primary consolidation with significant settlements to come. The situation is thus problematic for an airport island.

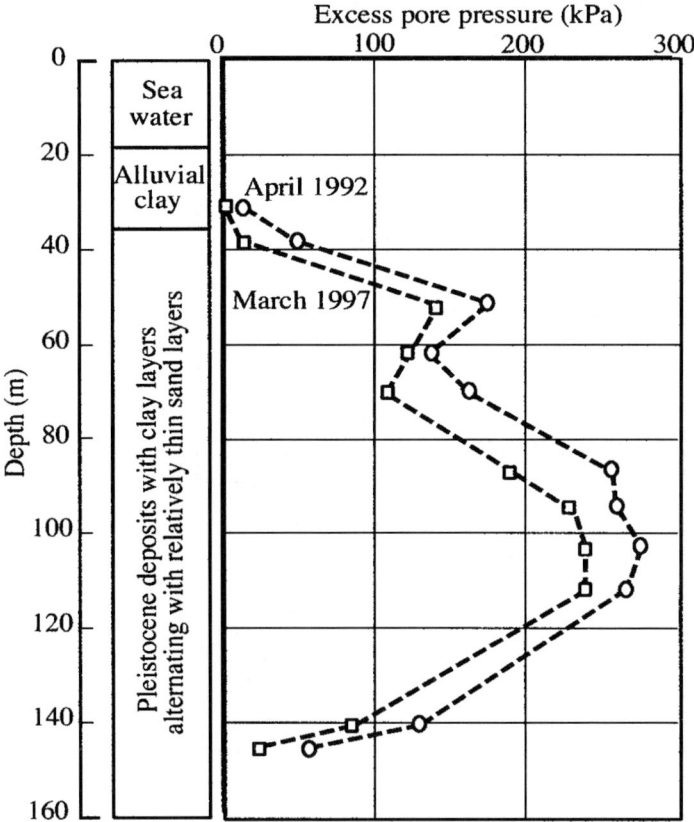

Figure 2 Excess pore pressure below Kansai International Airport Island (monitoring location N° 4; after Akai and Tanaka, 1999; Akai, 2000)

It is concluded from the observations made at KIA that for the same vertical strain in the laboratory and *in situ*, the in situ pore pressure is larger than the pore pressure expected based on the laboratory compression curve. As indicated on Figure 3, this means in situ stress conditions at a point such as S rather than at a point such as L. In other words, the *in situ* effective stress-strain curve is below the laboratory curve. However, as shown in Figure 4 and elsewhere in the literature, the same type of behaviour has been observed for most well-documented embankments on soft clays (Larsson, 1986; Kabbaj *et al.*, 1988; Leroueil, 1988, 1996 and 2001) and is attributed to the viscous behaviour of clays. As the strain rates *in situ* are typically 100 to 10,000 times smaller than in conventional laboratory tests, the *in situ* effective stress-strain curve lies below the one found in the laboratory. At KIA where the *in situ* strain rate is more than 3 orders of magnitude smaller than the one in conventional laboratory tests, settlements larger than those calculated on the basis of these latter tests can be expected.

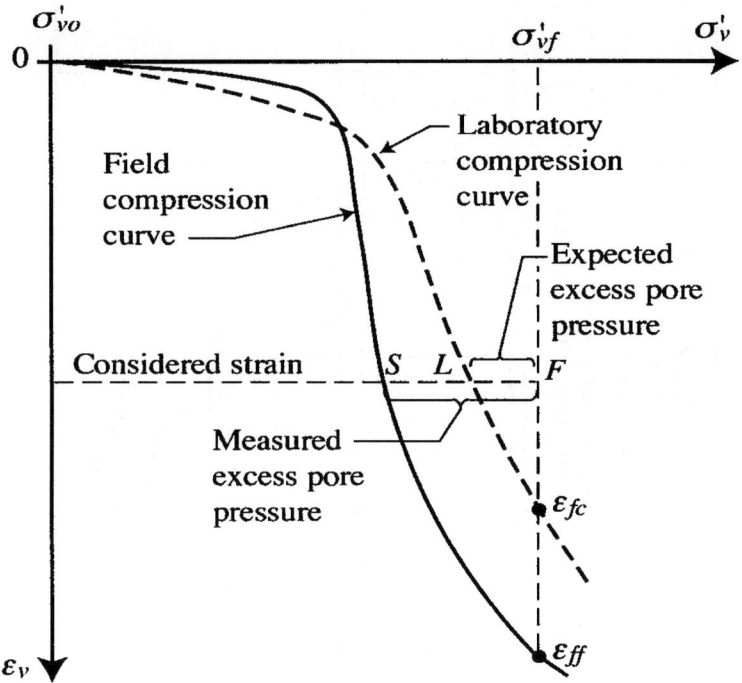

Figure 3 Typical compression curves both *in situ* and in the laboratory

Is Kansai Pleistocene clay problematic? The author does not think so, as the Pleistocene clay seems to behave as all other clays.

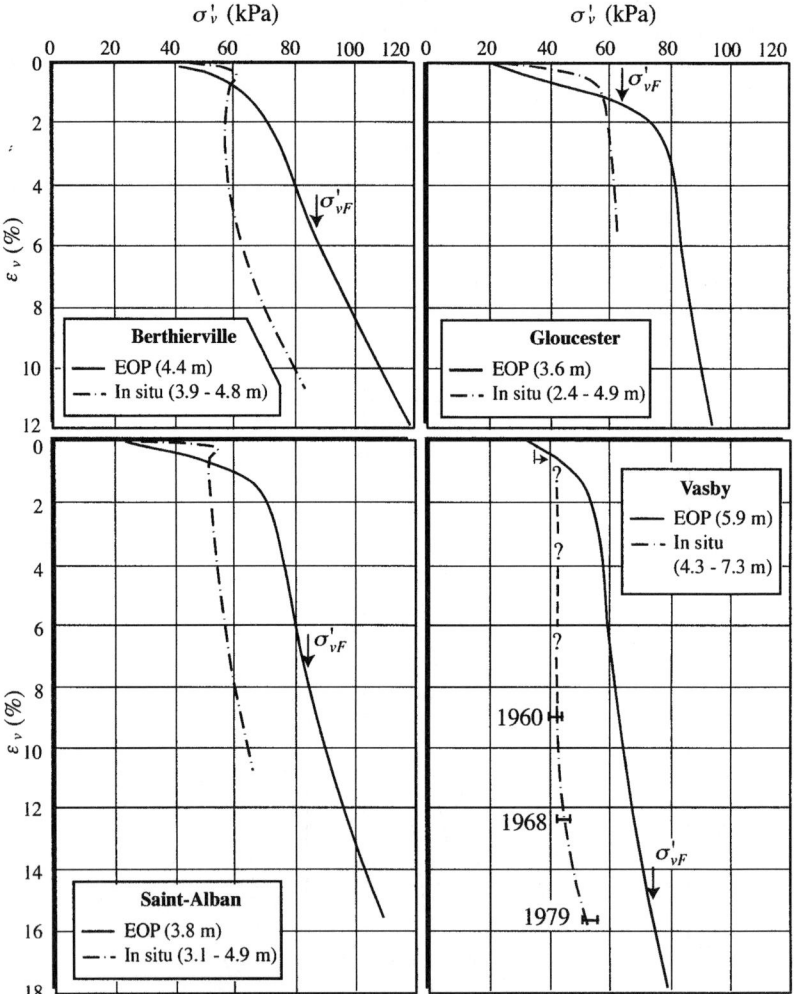

Figure 4 Comparison between stress-strain relations observed *in situ* and measured at end-of-consolidation in the laboratory on four test sites (after Kabaj *et al.*, 1988)

Transient water flow through earth dams

Observation of pore pressure distribution in the core of several earth dams has shown unexpected behaviour with pore pressures in the downstream portion of the core much higher than those corresponding to the distribution obtained assuming a constant hydraulic conductivity. There is then a concentration of equipotential lines in the downstream portion of the core (Dascal, 1984; Verma *et al.*, 1985; Vaughan, 1989 and 1994; Stewart *et al.*, 1990; Peck, 1990; Kleiner, 1997; Alicescu, 2000; Alicescu *et al.,* 2000; Sobkowicz *et al.*, 2000). Figures 5 and 6 show two Canadian cases, the LG-4 main dam from Quebec and the WAC Bennett Dam from British Columbia, which have shown such pore pressure distributions. Figure 7 also shows the variation of pore pressures with time in some piezometers located in the core of the WAC Bennett Dam. It can be seen that the pore pressures increase for several years after the filling of the reservoir, reach a peak and then progressively decrease.

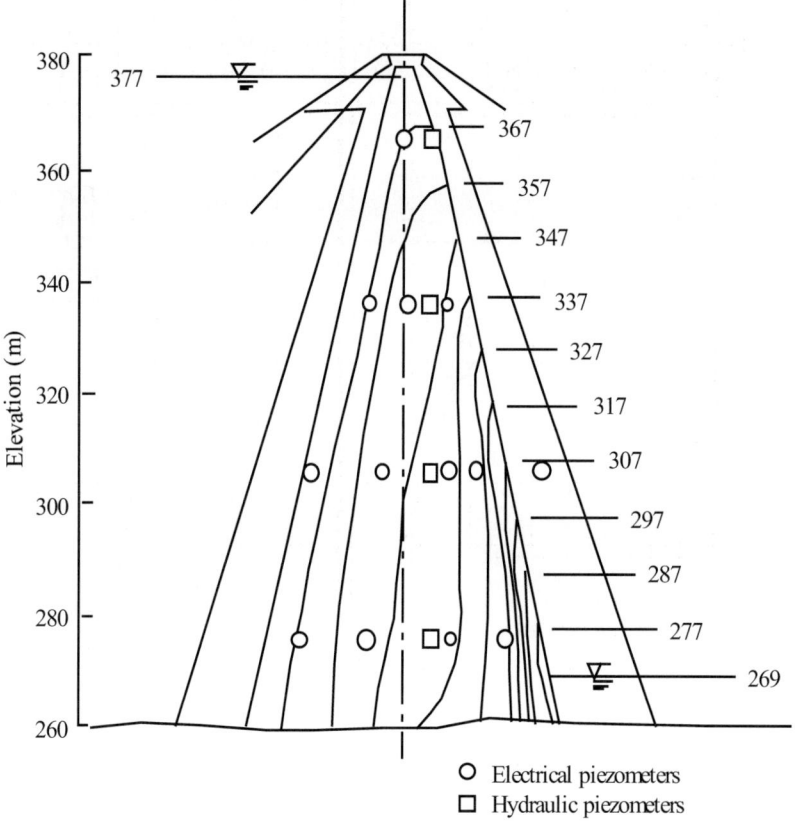

Figure 5 LG-4 Main Dam. Equipotential lines at the end of reservoir filling after Verma et al., 1985)

Such behaviour has been considered problematic because it would decrease the factor of safety against failure of the dam, but mostly because it was not clearly understood. Several hypotheses have been proposed to explain this behaviour: Dascal (1984) suggests the existence of zones of weakness that could have originated from horizontal cracks due to arching; Sherard (1986) emphasizes that some fine particles can move in the core into cracks initiated by hydraulic fracturing; Vaughan (1989, 1994) explains the apparent lower hydraulic conductivity of the downstream portion of some cores by the fact that

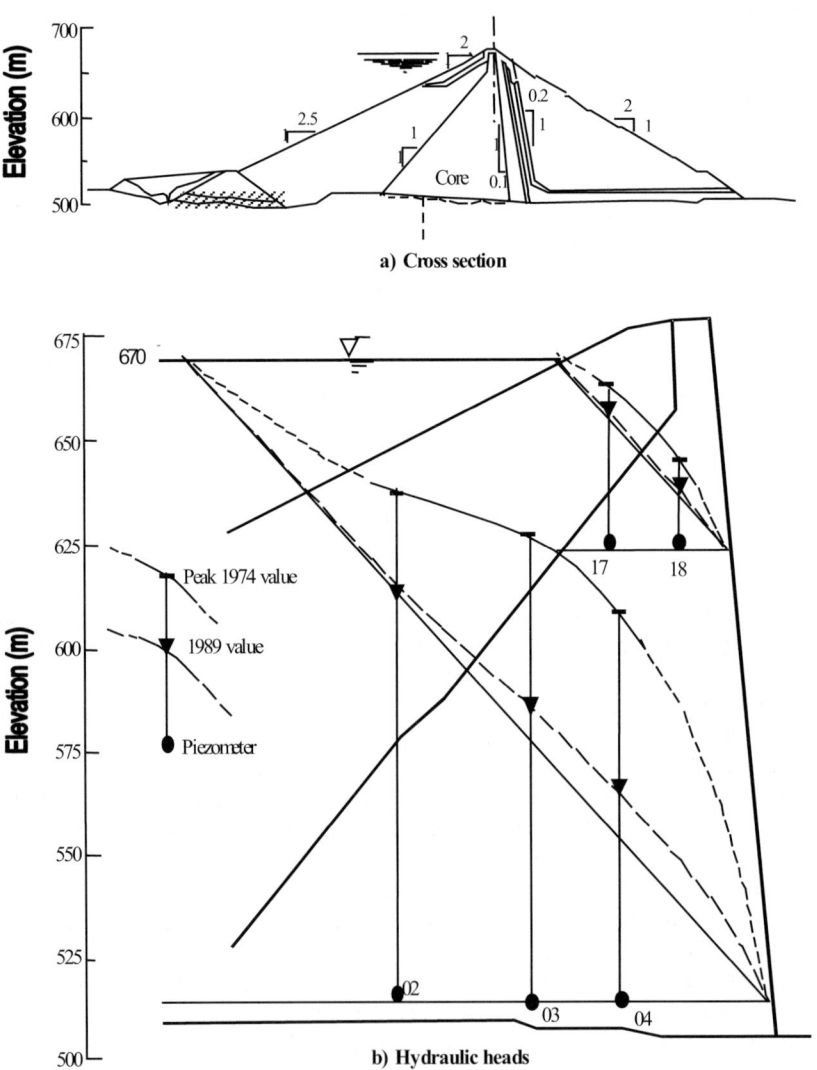

Figure 6 WAC Bennett Dam a) cross section, b) hydraulic heads in piezometers (after Stewart *et al.*, 1990)

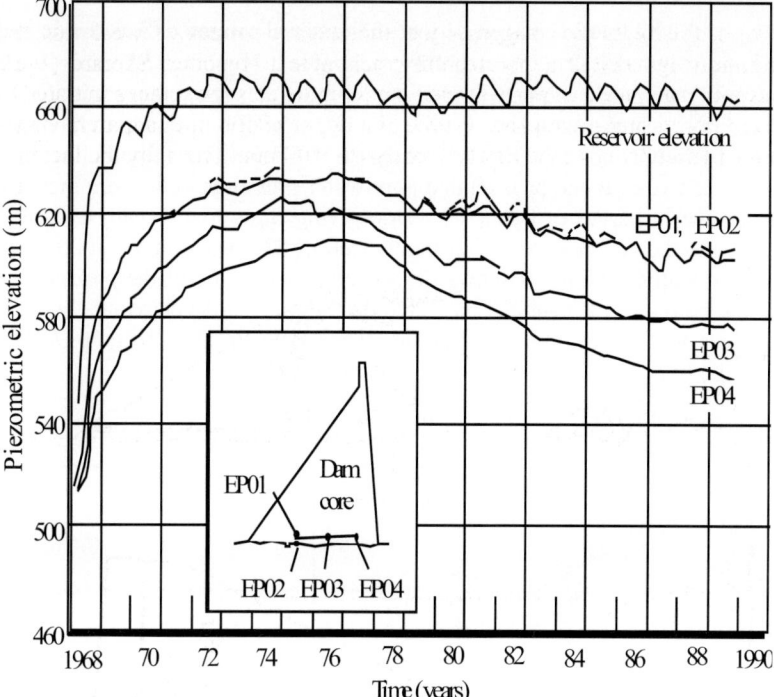

Figure 7 Typical piezometric pressure response with time. WAC
Bennett Dam (after Stewart *et al.*, 1990)

the vertical effective stress there is higher and the void ratio smaller; Peck
(1990) suggests that fine particles of the core may migrate to the downstream
filter, clog it and accumulate into a barrier.

Cracks, hydraulic fracturing and movement of fines do not sound well when
talking about dams. St-Arnaud (1995) suggests a completely different
explanation based on the idea that some air is trapped in the compacted core
during reservoir impounding. This air would be compressed and partially
dissolved by water in the upstream part of the core; the dissolved air would be
transported by water flowing through the core and would come out of solution
downstream as water pressure decreases. All these processes would induce
variations of the hydraulic conductivity, depending on the position of the soil
element in the core and time.

Le Bihan and Leroueil (2001) examined numerically the practical
implications of the hypothesis put forward by St-Arnaud (1995), using the
fundamental laws of Boyle, Henry, Fick and Darcy, and a hydraulic
conductivity:

$$k = k_{sat} \, S_r^{\alpha k}, \tag{1}$$

where k_{sat} is the hydraulic conductivity of the material saturated, S_r is the degree of saturation and αk is a permeability parameter. Hypothetical cases were studied and parametric studies carried out. Figure 8 shows the evolution of calculated pore water pressure with time at a depth of 100 m in a core having a saturated hydraulic conductivity k_{sat} equal to 10^{-7}m/s. The impounding was made over one year. It can be seen that pore water pressures in the core increase with time over a period of about 10 years, giving high downstream hydraulic gradients, and then decrease to reach steady state conditions after about 25 years for the case studied. Figure 9 shows the variation with time of the piezometric head at the points B1, B2 and B3 indicated on Figure 8. The behaviour shown is very similar to the one observed at the WAC Bennett Dam (Figure 7) and validates the hypothesis put forward by St-Arnaud (1995).

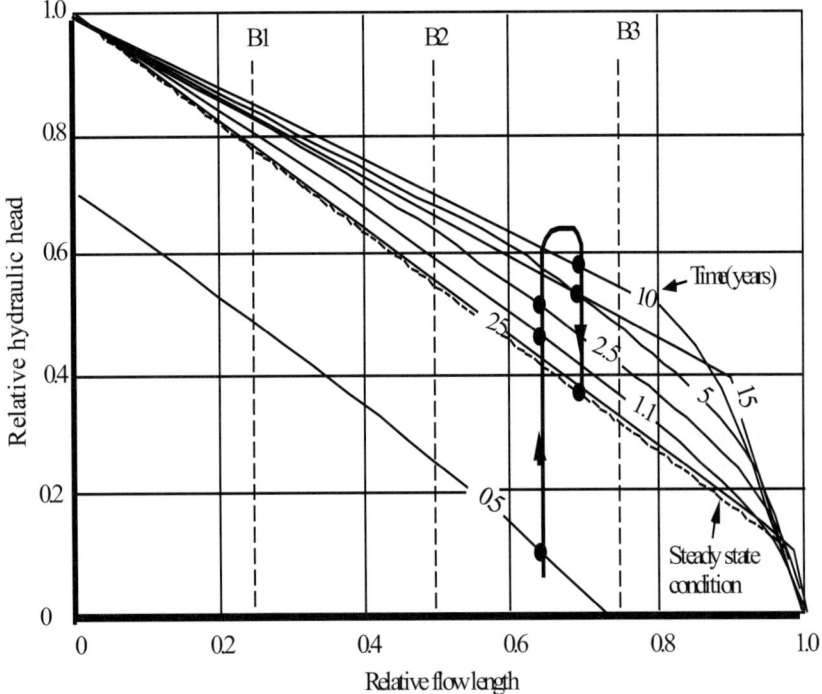

Figure 8 Relative head data at a depth of 100m as a function of relative flow length for different times (after Le Bihan and Leroueil, 2001)

The numerical program developed by Le Bihan and Leroueil (2001) was used to simulate the behaviour observed in the LG-4 Dam (Figure 5). Impounding of the reservoir was completed by the end of 1983. With permeability characteristics very close to those defined in the laboratory, the calculated and measured pore water pressures are very close both in space and time (Alicescu *et al.*, 2000). A comparative example is shown in Figure 10 for level 276m (see Figure 5) at an average depth of about 98m below the level of the reservoir. Pore water pressures first increase to reach their maximum around 1990 and then progressively decrease. These results indicate that the unexpected elevated

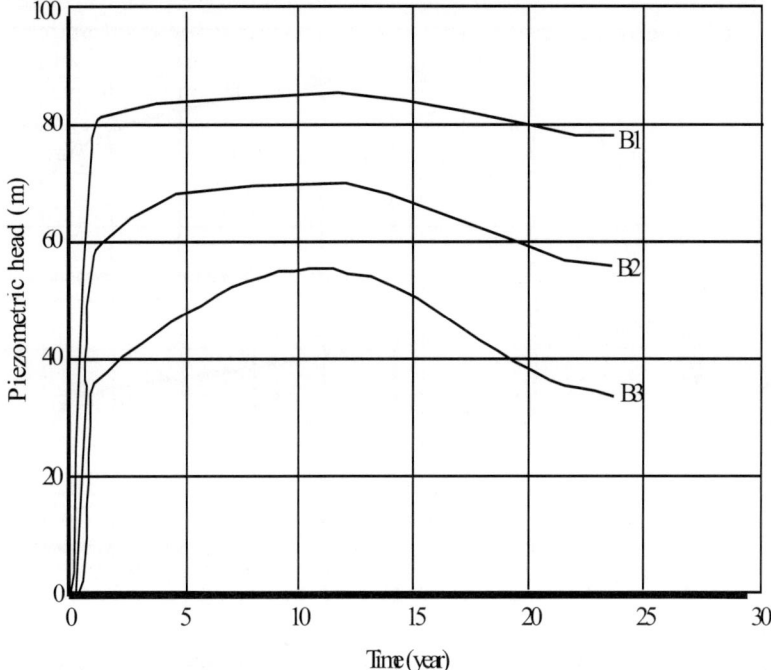

Figure 9 Evaluation of the piezometric head with time data depth of 100m (after Le Bihan and Leroueil, 2001)

pore water pressures result from the fundamental laws of physics governing the flow through an initially unsaturated core and not from the soil itself. This, of course, does not change the observed phenomena but indicates that the unexpected pore pressures are not associated with abnormal behaviour in the earthworks such as might arise from arching effects, hydraulic fracturing, or movement of fines through the core.

Specifications for the compaction of soils used as hydraulic barriers

Benson *et al.,* (1999) examined a database consisting of 85 full-scale compacted clay liners and field test pads for evaluating performance based on field hydraulic conductivity. Nearly all regulations for clay liners in the United States require that the hydraulic conductivity does not exceed 10^{-9} m/s. The database indicates that 26% of the clay liners failed this requirement, even if 92% of the corresponding laboratory tests were showing hydraulic conductivities between 2×10^{-11} m/s and 5×10^{-10} m/s. In most cases the primary cause was that compaction was made dry of the optimum moisture content. A similar situation seems to exist in Canada (Leroueil *et al.*, 1990). This suggests that constructing clay liners to this level of impermeability may be problematic.

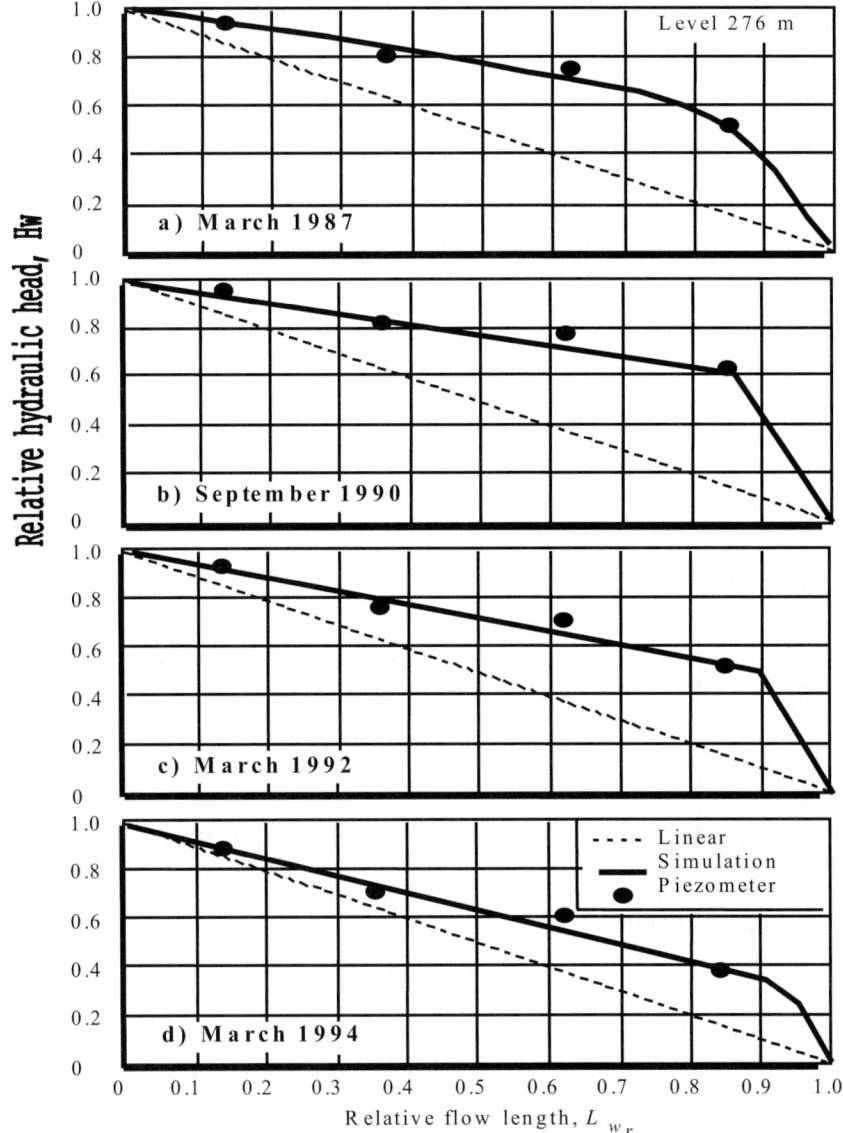

Figure 10 Measured and calculated relative hydraulic heads at level 276 m in the core of LG-4 dam, at different times (from Alicescu *et al.*, 2000)

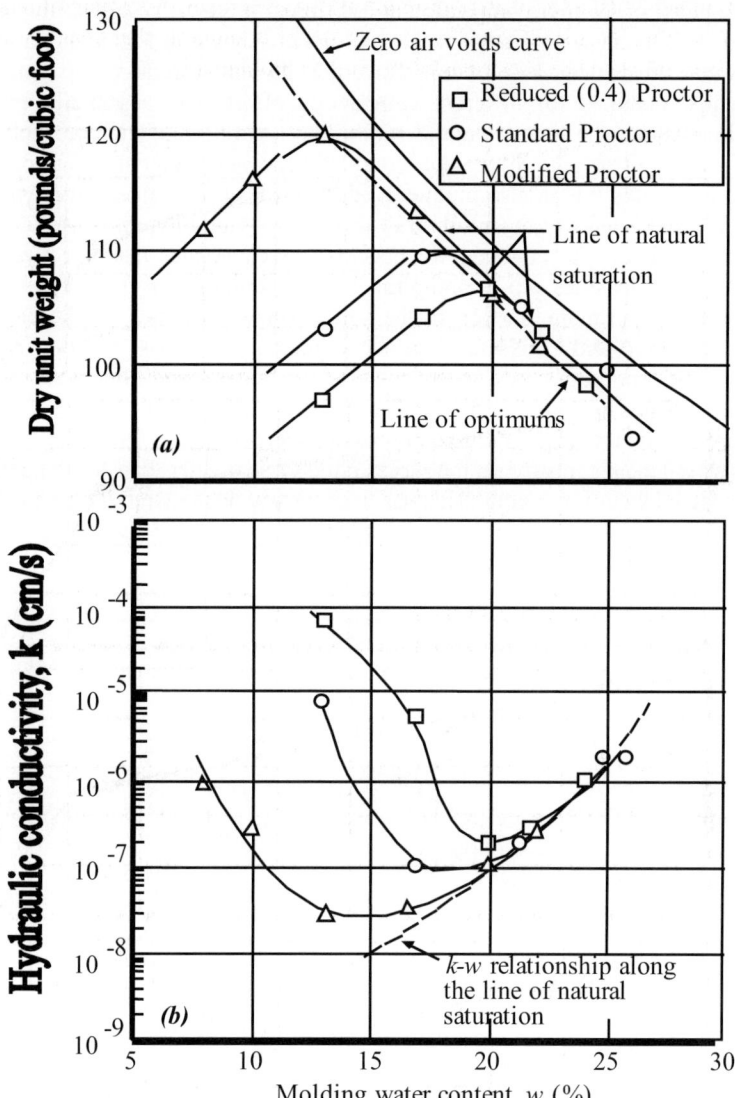

Figure 11 Example of compaction curves (a) and hydraulic conductivity versus molding water content relationships (b) (experimental data from Daniel and Benson, 1990)

It has been recognized for about 40 years (Bjerrum and Huder, 1957; Mitchell *et al.*, 1965) that compaction conditions influence hydraulic conductivity. A good example is provided by Daniel and Benson (1990). These authors measured the hydraulic conductivity of a saturated clayey soil having a plasticity index of 19 after compaction at 3 different levels of energy of the soil prepared at different water contents. The results are shown in Figure 11. It can be seen that on the wet side of optimum, the dry unit weight-water content relationships obtained for different compaction efforts, are essentially on a unique line, called the line of natural saturation, regardless of the compaction effort (Figure 11a). The hydraulic conductivity-water content relationship obtained along this line is also unique, (Figure 11b). On the other hand, when the soil is compacted dry of optimum at degrees of saturation less than at the optimum, the hydraulic conductivity becomes much greater than for similar compaction dry densities wet of optimum. This is explained by the fact that, on the wet side of optimum, suction in the soil is small and clods can be easily broken, deformed and compressed into a homogeneous mass. On the other hand, dry of optimum, suction in the soil is higher and the clods less deformable. Consequently, a bimodal pore size distribution develops with micropores inside the clods and macropores between the clods. As the hydraulic conductivity depends mostly on the larger pores, it is greater than for the same soil compacted wet of optimum. This has been clearly demonstrated for clays; it can also be true for non-plastic silty materials, as shown on Figure 12 from Watabe *et al.*, (2000).

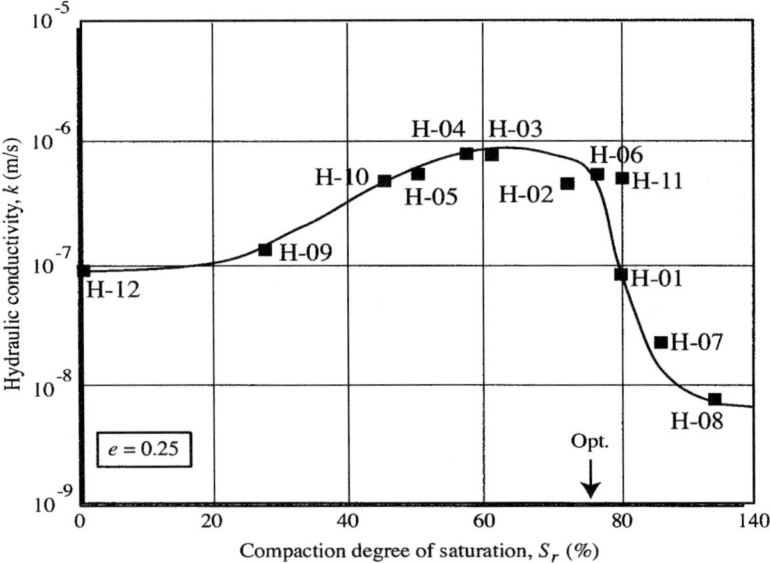

Figure 12 Hydraulic conductivity as a function of the compaction degree of saturation at a void ratio of 0.25, LG-2 till (after Watabe *et al.*, 2000)

Figure 12 shows that at the same void ratio of 0.25, the hydraulic conductivity can vary from about 10^{-10} m/s when compacted at degrees of saturation larger than the one at the optimum (75% for this glacial till) to about 10^{-8} m/s for compaction degrees of saturation between 50% and 70%.

To have a low hydraulic conductivity, the compacted soil must have a homogeneous fabric and, for that purpose, be compacted at degrees of saturation larger than the one at the optimum. Construction specifications must reflect this fact and should not be based on a minimum percent compaction, as is often stipulated. Leroueil *et al.,* (1992) suggests the specification shown in Figure 13 and stated that for clay liners: the clayey soil must be compacted at degrees of saturation equal to or larger than the one on the line of optimums; the compaction water content must be such that the undrained shear strength is small enough so that the clods can be easily broken, deformed and compressed yet large enough so that the compaction equipments will not produce excessive rutting (limit of trafficability); the lower limit for the compaction water content has to be defined in order to avoid the development of fissures in the soil. Benson *et al.,* (1999) has proposed similar criteria.

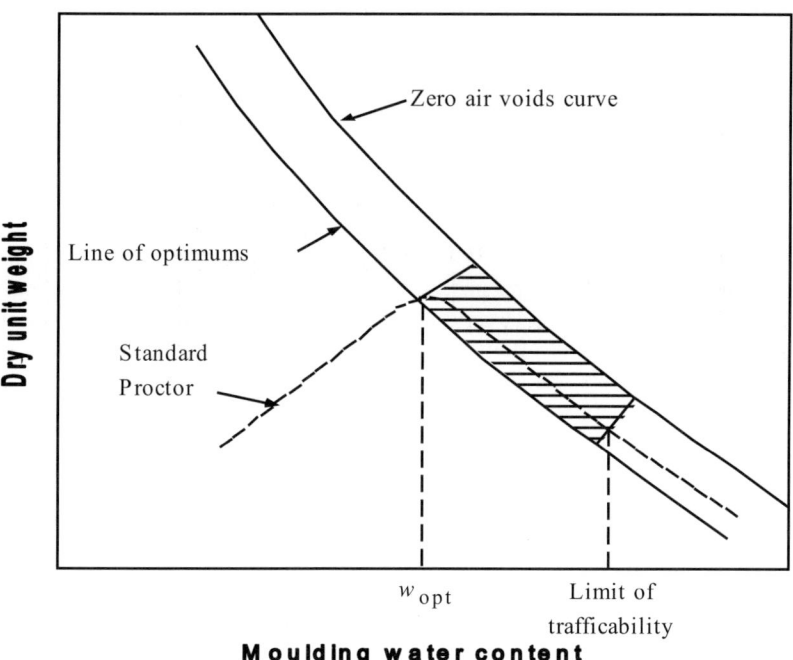

Figure 13 Recommended compaction zone for clay liner (from Leroueil *et al.*, 1992)

This case shows that a good understanding of the mechanisms controlling the fabric of compacted soils leads to specifications that are well adapted for the construction of liners in which low hydraulic conductivity is required. In a paper to this symposium, Coombs *et al.*, (2001), give another example where specifications were apparently not well adapted to the use of recycled materials such as Pulverised Fuel Ash. In both cases, it is because the specification was not based on the parameter controlling the performance or on the controlling factors.

Risk assessment and reliability-based analysis

There are situations for which the problem cannot be completely eliminated by the OUT approach. This is particularly the case for natural hazards such as earthquakes or landslides. It is, however, the role of our profession to minimize the risk associated with the phenomenon as much as possible, and risk assessment or reliability-based analysis appears to be a rational approach to such problems.

Varnes *et al.* (1984) defined the risk as follows:

$$R = H \sum_i E_i V_i \tag{2}$$

in which H is the hazard or the phenomenon occurrence probability within a given period of time; E_i (for $i = 1$ to n) are the elements at risk, potentially damaged by the phenomenon; and V_i is the vulnerability of each element represented by a damage degree number between 0 (no loss) and 1 (total loss). The elements at risk E_i can be individuals, properties and goods, activities and social functions. $\sum_i E_i V_i$ represents the total cost of damages.

In some cases, such as earthquakes, hazard cannot be reduced, and the only way to reduce the risk is to decrease the consequences of the event. This can be done, for example, by reducing the liquefaction potential of sand layers, strengthening buildings, etc. In other cases, such as landslides, the risk can be minimized by reducing hazard and/or the consequences. In all cases, the reduction in risk will be obtained by choosing an appropriate construction site, selecting appropriate design methods or using mitigating techniques. In all cases, a good understanding of the situation is absolutely essential for selecting an appropriate and adequate solution. It is, however, important to mention that the final solution may not be chosen on the basis of technical criteria alone, but also on the basis of social, economic, environmental, political and legal considerations. Societal acceptance is an additional factor. For example, there are many people in the world who are consciously living in areas with a significant probability of being affected by floods, landslides or volcanic eruptions.

An approach to risk assessment and mitigation of slope movements has been proposed by Leroueil and Locat (1998). This approach is based on the geotechnical characterization of slope movements suggested by Leroueil *et al.* (1996). This characterization takes into account the type of material and the type of movement involved as well as the stage of movement: pre-failure, failure, post-failure or reactivation. For each relevant case, a characterization sheet is established with the following: a) the controlling laws and parameters; b) the predisposition factors; c) the triggering or aggravating factors; d) the revealing factors; and e) the consequences of the movement (Figure 14). This geotechnical characterization provides a rational and useful framework for risk assessment as the hazard is directly the probability of the triggering factor reaching a critical value and leading to failure. The elements at risk and their vulnerability should be, directly or indirectly, in "Movement consequences" (Figure 14).

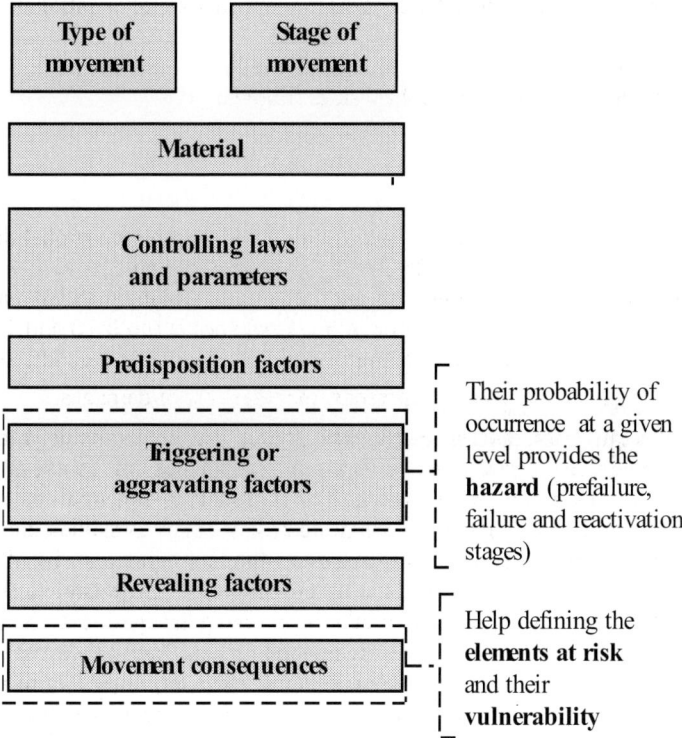

Figure 14 Geotechnical characterization and risk assessment (after Leroueil and Locat, 1998)

Bell and Culshaw (2001) provide an example where technical and economical aspects intervene and to some extent are in opposition. They estimate £3 billion losses have been caused by swelling and shrinkage in clay soils in Britain and between US$6 billion and US$11 billion costs annually for the damage to property associated with expansive soils in the United States. Driscoll and Chown (2001) provide further examples of this. The phenomena involved are relatively well understood, and testing techniques and design approaches have been developed to avoid the problem or to limit it to an acceptable level. However, applying these techniques requires a good understanding of the problem and generally increases construction costs by an amount that could exceed the potential cost of damages. This problem thus becomes economic rather than technical and should be examined on a global cost-benefit basis. This type of approach has been successfully applied to the treatment of contaminated land. Bunce and Braithwaite (2001) provide an excellent example of this in this symposium.

Combined with a good understanding of the phenomena involved and appropriate technologies, risk assessment or reliability-based analysis thus appears to provide a rational approach to assist in decision-making processes and minimizing the problems associated with a given situation.

Conclusions

The author does not think that there are problematic soils. He is convinced, however, that there may be problematic understanding of soil behaviour, problematic engineers, problematic designs, problematic specifications, problematic uses of technologies, problematic situations, a problematic shortage of money, a problematic lack of political will, etc. The main role of the geotechnical engineer is certainly to try to mitigate these problems. Observation, improvement of our understanding and technological development (the OUT approach) leads to a logical framework for problem solving. More specifically:

-   We have to improve our knowledge of soil behaviour in order to tackle geotechnical problems with more confidence. It is thought that the approach advocated by the author (Leroueil, 1997) and others, aiming at extending the concepts of limit and critical states to include the influence of anisotropy, discontinuities, viscosity, microstructure and partial saturation, provides a general and useful framework that is applicable to most natural geomaterials.
-   This improved understanding must be reflected in the way engineers think about soil behaviour, develop designs and apply available technologies in their projects. However, a major difficulty appears to be the transfer of knowledge from researchers to practitioners.
-   More effort should be invested in developing and improving technologies that aim at mitigating problems.
-   Risk assessment or reliability-based analysis seems to provide a rational approach to assist in decision-making processes and to minimize the consequences of a potential problem.

References

Akai, K. 2000. *Insidious settlement of super-reclaimed offshore seabed.* Int. Symp. on Coastal Geotechnical Engineering in Practice, Yokohama, 1, 243-248.

Akai, K. & Tanaka, Y. 1999. *Settlement behaviour of an offshore airport KIA.* 12th European Conf. on Soil Mechanics and Geotechnical Engng., Amsterdam, 2, 1041-1046.

Alicescu, V. 2000. *Particular behavior of Robert Bourassa Main Dam, La Grande-2 Complex, James Bay,* 53rd Canadian Geotechnical Conf., Montréal, 1, 219-226.

Alicescu, V., Le Bihan, J.P. & Leroueil, S. 2000). *Simulation of transient flow through the LG-4 earth dam, Quebec.* 53rd Canadian Geotechnical Conf., Montréal, 1, 203-209.

Arai, Y., Oikawa, K. & Yamagata, N.1991. *Large-scale sand drain works for the Kansai International Airport Island.* Int. Symp. Geo-Coast'91, Yokohama, 1, 281-286.

Bell, F. G., & Cushaw, M.G. 2001. *Problem soils: a review from a British perspective.* In Problematic Soils Symp. (Jefferson, I., Murray, E. J., Faragher, E., and Flemming, P. R. eds), Nottingham, *(in press).*

Benson, C.H., Daniel, D.E. & Boutwell, G.P. 1999. *Field performance of compacted clay liners.* J. of Geotechnical and Geoenvironmental Engng., ASCE, 125, (5), 390-403.

Bjerrum, L. & Huder, J. 1957. *Measurement of the permeability of compacted clays.* 4th Int. Conf. on Soil Mechanics and Foundation Engng., London, 1: 6-10.

Bunce, D. & Braithwaite, P. 2001. *Reclamation of contaminated land with specific reference to Pride Park: Derby.* In Problematic Soils Symp. (Jefferson, I., Murray, E. J., Faragher, E., and Flemming, P. R. eds), Nottingham, *(in press).*

Coombs, R., Sear, L. K. A. & Weatherly, A. 2001. *Problematic soils or is it problematic specifications?.* In Problematic Soils Symp. (Jefferson, I., Murray, E. J., Faragher, E., and Flemming, P. R. eds), Nottingham, *(in press).*

Daniel, D.E. & Benson, C.H. 1990. *Water content-density criteria for compacted soil liners.* J. of Geotechnical Engng., ASCE, 116, 1811-1830.

Dascal, O. 1984. *Peculiar behaviour of the Manicouagan 3 dam's core.* Int. Conf. on Case Histories in Geotechnical Engng., St-Louis, 2, 561-567.

Driscoll, R. & Chown, R. 2001. *Shrinking and swelling of clays.* In Problematic Soils Symp. (Jefferson, I., Murray, E. J., Faragher, E., and Flemming, P. R. eds), Nottingham, *(in press).*

Kabbaj, M., Tavenas, F. & Leroueil, S. 1988. *In situ and laboratory stress-strain relations.* Géotechnique, 38(1), 83-100.

Kleiner, D.E. (1997. Pore pressure response in dams subject to rapid filling and emptying – Rocky Mountain Project. 19th Congress on Large Dams, Florence. Q.73, R.9, 123-144.

Larsson, R. 1986. Consolidation of soft soils. Swedish Geotechnical Institute, Report No. 29, Linköping, Sweden.

Le Bihan, J.P. & Leroueil, S. 2001. A theoretical model for gas and water flow through the core of earth dams. Canadian Geotechnical J. *(in press).*

Leroueil, S. 1988. Recent developments in consolidation of natural clays. Canadian Geotechnical J., 25, (1), 85-107.

Leroueil, S. 1996. Compressibility of clays: fundamental and practical aspects. J. of the Geotechnical Engng., ASCE, 122, (7), 534-543.

Leroueil, S. 1997. Critical state soil mechanics and the behaviour of real soils. Int. Symp. on Recent Developments in Soil and Pavement Mechanics, Rio de Janeiro, 41-80.

Leroueil, S. 2001. *Some fundamental aspects of soft clay behaviour and practical implications.* 3rd Int. Conf. On soft soil, Engng., Honk Kong, *(in press)*

Leroueil, S. & Vaughan, P.R. 1990. *The general and congruent effects of structure in natural soils and weak rocks.* Géotechnique, 40, (3), 467-488.

Leroueil, S. & Locat, J. 1998. *Slope movements – Geotechnical characterization, risk assessment and mitigation.* 8th Int. IAEG Congress, Vancouver, 2, 933-944.

Leroueil, S., Bouchard, R. & Bourret, M. 1990. *Influence des conditions de mise en place sur la performance d'une membrane d'argile.* 43rd Canadian Geotechnical Conf., Québec, 1, 369-375.

Leroueil, S., Le Bihan, J.-P. & Bouchard, R. 1992. *Remarks on the design of clay liners used in lagoons as hydraulic barriers.* Canadian Geotechnical J. 29, (3), 512-515.

Leroueil, S., Vaunat, J., Picarelli, L., Locat, J., Lee, H. & Faure, R. 1996. *Geotechnical characterization of slope movements.* 7th Int. Symp. on Landslides, Trondheim, 1, 53-74.

Mitchell, J.K., Hooper, D.R. & Campanella, R.G. 1965. *Permeability of compacted clays.* J. Soil Mechanics and Foundation Div., ASCE, 91, (SM4), 41-65.

Peck, R.B. 1990. *Interface between core and downstream filter.* H. Bolton Seed Memorial Symp., 2, 237-251.

Sherard, J.L. 1986. *Hydraulic fracturing in embankment dam.* J. Geotechnical Engng. Div., ASCE, 112, (10), 905-927.

Sobkowicz, J., Byrne, P., Leroueil, S. & Garner, S. 2000. *The effect of dissolved air and free air on the pore pressures within the core of the WAC Bennett Dam.* 53rd Canadian Geotechnical Conf., Montréal, 1, 87-95.

St-Arnaud G. 1995. *The high pore pressures within embankment dams: an unsaturated soil approach.* Canadian Geotechnical J., 32(6): 892-898.

Stewart, R.A., Imrie, A.S. & Hawson, H.H. 1990. *Unusual behaviour of the core at WAC Bennett Dam.* 43rd Canadian Geotechnical Conf., Quebec, 2, 549-558.

Varnes, D.J. & the IAEG Commission on Landslides and Other Mass Movements on Slopes. 1984. *Landslide Hazard Zonation – A Review of the Principle and Practice*. UNESCO, Paris.

Vaughan, P.R. 1989. *Non linearity in seepage problems – Theory and field observation*. De Mello Volume, Bitech Publisher Ltd., 501-516.

Vaughan, P.R. 1994. *Assumption, prediction and reality in geotechnical engineering*. Géotechnique, 44, 573-609.

Vaughan, P.R. 1999. *Problematic soil or problematic soil mechanics*? Int. Symp. on Problematic Soils, IS-Tohoku'98, Sendai, 803-814.

Verma, N.S., Paré, J.-J., Boncompain, B. Garneau, R. & Rattue, A. 1985. *Behaviour of the LG-4 main dam*. 11th Int. Conf. on Soil Mechanics and Foundation Engng., San Francisco, 4, 2049-2054.

Watabe, Y., Leroueil, S. & Le Bihan, J.P. 2000. *Influence of compaction conditions on pore size distribution and saturated hydraulic conductivity of a glacial till*. Canadian Geotechnical J. 37, (6), 1184-1194.